MECHANICAL ENGINEERING SERIES

EDITORS

S. Paolucci   M. Sen

Undergraduate Lectures on Statics
Dynamics
CAD/Engineering Graphics
Measurements and Data Analysis
Solid Mechanics
Mechanical Vibrations
Control Systems
Kinematics
Machine Design
Thermodynamics
Intermediate Thermodynamics
Fluid Mechanics
Heat Transfer

The books in this Mechanical Engineering Series educate readers on critical subjects included in undergraduate mechanical engineering curricula. Each topical volume is concise, written specifically for students new to the material, and complete in presenting the essentials covered in a one term course. The books are highly readable in style, rigorous in detail of underlying concepts, and emphasize the basic principles needed for engineering applications. The chapters of each volume are discrete units that instructors can present as individual lessons and students can use to accompany the lectures.

D1531022

P. Rumbach
University of Notre Dame

Undergraduate
Lectures on
Measurements and
Data Analysis

BreviLiber

First published 2021

Library of Congress Cataloguing in Publication data
Names: P. Rumbach, author.
Title: Undergraduate Lectures on Measurements and Data Analysis/ P. Rumbach, University of Notre Dame.
Description: South Bend, IN : BreviLiber, 2021. | © 2021| Includes illustrations, examples and index.

Typeset in LaTeX
Printed by KDP

ISBN 979-8-534-42463-8 Paperback

Series Preface

Commonly used engineering textbooks have usually undergone through many editions since they were first written by the primary author(s). They have usually grown considerably in size; from perhaps 300 pages in the first edition to three times as many in a current edition. Books just seem to grow while the number of hours spent in the classroom has not. The real downside is that such lengthy books are not read as much as the authors expect or hope. Indeed, much of the contents of a typical textbook are not even included in a course but are there for possible future reference. A student who is just learning the material quite often has difficulty identifying the "wheat" from the "chaff" in such voluminous tomes. Furthermore, it is common to have a large number of unsolved problems at the end of each chapter, and physical property tables and charts that take up considerable space at the end of a book. This leads to a culture of not reading the material that is essential for a course.

This volume is part of a series of books that are disposable, inexpensive and brief, and ones that the student should read, write notes on, and dog-ear to get an understanding of the subject matter. These books are written for the student rather than for the instructor, and are intended to be used in conjunction with in-class lectures. The material covered in each book is presented as a series of lectures, where each lecture includes at least one example. As such, we hope that instructors will also find the present book useful in the classroom. The content of each lecture can actually be covered in a standard lecture period. The books in the Series do not have lots of examples and a long list of unsolved exercises or problems at the end of each chapter. Neither do the books have a list of physical properties of materials. Solved and/or unsolved problems can be easily obtained from other sources for classroom purposes (there are even books in many areas that specialize entirely in problems). In addition, most physical properties are easily obtainable from the Internet (as an example, for the thermodynamic properties of water one can download an open-source code which will do the job). Such searches enhance the students' learning since actual learning is an active and not a passive process. Integrals, trigonometric identities and similar information that can be obtained from the Internet are also eliminated.

Lastly, and to be clear, books in this Series are not watered down in terms of the level of physics and mathematics used in an engineering curriculum and the effort needed by the student to master the topics; no change is expected in either of those fronts from what is currently expected from the student.

South Bend, Indiana   S. PAOLUCCI AND M. SEN

Preface

Good engineering decisions are based on scientific laws and logical interpretations of measured data, and data-driven decision making has become the de facto standard in the business world. Additionally, most mechanical products, both consumer and industrial, now contain with a whole suite of electronic sensors and circuits. Thus, the following learning objectives should be of paramount importance to any engineering student.

After reading this book, a student should be able to:

1. Know the working principles behind a variety of transducers and sensors.

2. Confidently choose electronic sensors and design tertiary circuits and digital data acquisition systems.

3. Process and present measured data in a professional manner using Matlab.

4. Understand the limitations of various sensors and measurement techniques through the use of uncertainty calculations.

5. Analyze and assess experimental results and data sets using the basic mathematical tools of probability and statistics.

The scope of this book is limited to only *fundamental* skills and techniques likely to be taught in a sophomore or junior-level engineering laboratory course on measurements and data analysis. If any particular topic (digital electronics, probability, statistics, etc.) arouses your interest, I strongly encourage you to seek out an upper-level elective course, where you can gain a deeper understanding.

Notre Dame, Indiana   P. Rumbach

Dedicated to my Father, David A. Rumbach, Notre Dame class of 1980.

Contents

1 Introduction

Billions of people have lived in our world, and each one has their own unique set of experiences and observations. Based on these experiences, every person forms a unique set of beliefs and opinions. As collegiate scholars, our overall goal is to determine the **truth** and distinguish fact from opinion. As engineers, our overall goal is to design and build products that work well and are safe for people to use. Designs based on a scientific facts are far more reliable and safe than ones based on opinions and intuition.

Notes: Throughout this book, keywords will be marked in **bold**. Students should strive to understand and internalize every keyword.

1.1 The Scientific Method

How do we distinguish lies from **truth**, opinions from facts? As engineers, we use the **scientific method** . The scientific method transcends the subjective experience of a single individual by using objective data collected by multiple individuals. This is an important distinction—the scientific method relies on repeatable observations. That is, a trend or phenomenon must be observed and recorded by multiple individuals at different places and times for it to gain credibility.

The scientific method is based on a set of logical steps:

1. Ask a **scientific question** about a given subject.

2. **Read** what others have written about the subject.

3. Formulate a logical **hypothesis** based on your reading.

4. Design and conduct an **experiment** to test the hypothesis.

5. Process and **analyze data** from the experiment.

6. **Communicate the results** through technical writing.

Notes: "Remember, kids: The only difference between screwing around and science is writing it down." – Adam Savage, *Mythbusters*

Note that the scientific method relies on the collection of data and observations from many different people. This is only achieved through steps 2 and 6: reading and writing technical literature.

1.2   Eratosthenes and the Radius of Earh

Over 2,200 years ago, an ancient Greek scholar named Eratosthenes used the scientific method to accurately determine the radius and circumference of the earth. Eratosthenes was the chief librarian at the famous library of Alexandria in North Africa. One particular text in the library described a phenomenon observed in the city of Syene, south of Alexandria. At noon on the summer solstice in late June, the sun would shine directly downward into the bottom of deep wells and vertical structures would cast no shadow. Meanwhile at the same time up north, the sun would still cast shadows. How was this possible?

Eratosthenes hypothesized that the earth was a sphere, and this logically explained the difference in shadows. Furthermore, he used geometry to analyze the phenomenon and calculate the radius of Earth. Shown in Fig. 1.1, parallel rays from the sun hit earth. The shadow cast by a vertical post in Alexandria forms and angle $\alpha = 0.122$radians, while the vertical post in Syene casts no shadow. Lines drawn from the vertical posts down into the ground intersect at the center of the Earth at an angle $\alpha = 0.122$ radians. (Both angles are congruent, because they are opposite interior angles of a parallel bisector.)

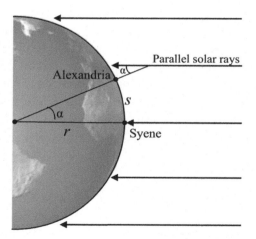

Figure 1.1: Eratosthenes' performed an experiment and used simple Euclidean geometry to determine the radius of Planet Earth.

The distance between Alexandria and Syene is $s = 800$km. Using the definition of an angle in radians $\alpha = s/r$, one can calculate the radius of the earth to be $r = 6,560$km. This is remarkably close to the modern accepted value of $6,371$ km!

Lastly, we only know that Eratosthenes performed this impressive measurenent, because *he took the time to write it down.*

1.3   Galileo's Inclined Plane

The scientific method establishes standard cause-and-effect relation-ships that can be applied universally. If a certain cause-and-effect re-lation is consistently observed under many different circumstances, it becomes a scientific law.

Scientific laws allow engineers to predict the behavior of a system before they even build it. A familiar example of this is Isaac Newton's Laws of Motion. The groundwork for Newton's laws of motion began nearly a century earlier with the experimental work of Galileo.

Among other things, Galileo was famous for his study of gravity, and he hypothesized that the motion of falling objects is governed by a set of mathematical rules. To test this hypothesis, he experimented by rolling balls down an inclined plane, as shown in Fig. 1.2. Specifically, he measured the time t that it takes for a ball to roll a distance x. His results showed that the motion was indeed governed by a specific mathematical rule: distance is proportional to the time squared, $x \propto t^2$.

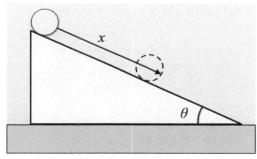

Figure 1.2: Galileo's famous experiment measured the time t it took a ball to roll a distance x down an inclined plane at angle θ.

The results from Galileo's inclined plane experiment ultimately lead to Isaac Newton's Laws of Motion. According to Newtonian mechan-ics, the distance travelled by the ball is related to time by the equation

$$x(t) = \frac{1}{2}\left(\frac{5}{7}g\sin\theta\right)t^2 + v_0 t + x_0 \tag{1.1}$$

where $g = 9.8$ m/s^2, θ is the angle of the inclined planed shown in Fig. 1.2, v_0 is the initial velocity, and x_0 is the initial position of the ball.

Example 1.1:

A cue ball is rolled down an inclined plane with an angle $\theta = 4.6°$, and the time t it takes the ball to roll a given distance x is measured.

distance, x (m)	time, t (s)
0	0
0.1	0.646
0.2	0.915
0.3	1.120
0.4	1.294
0.5	1.446

Using Matlab, plot the distance x as a function of the measured time t (i.e. distance on the vertical axis and time on the horizontal axis). Plot the theoretical trajectory, Eq. (1.1), as a continuous line over the data. Assume the ball is released from the initial position $x_0 = 0$ with initial velocity $v_0 = 0$. Include a legend to distinguish the measured data from the theoretical curve.

Matlab Script:

```
clc
close all

%measured data for cue ball (4.6 degrees)
x=[0 10 20  30  40  50]*0.01;  %meters
t=[0   0.646   0.915   1.120   1.294   1.446]; %seconds
theta=4.6;  %degrees

%theoretical data
|
g=9.8;  %m/s^2

tTheo=linspace(0,1.5,1000);     %seconds
xTheo=5/14*g*sind(theta)*tTheo.^2;  %meters

figure(1)
plot(t,x,'o',tTheo,xTheo,'LineWidth',1.5,'MarkerSize',8)
xlabel('time, t (s)')
ylabel('distance, x (m)')
axis([0 1.5 0 0.55])
set(gca,'FontSize',20)
legend('Measured data','Eq. (1.1)','Location','northwest');
```

Notes:

- The green text beginning with a % symbol is a comment that will not be executed as part of the code.

- Begin the scripts with 'clc' and 'close all' to clear the screen of clutter.

- Measured data can be hard-coded into the script in square brackets.

- Use 'linspace()' to create a vector for the independent variable in your theoretical equation.

- Every plot should be given a handle, such as 'figure(1)' shown here.

- Label the axes with units.

- The font should be large enough to read. Use a font size of 16 to 20 point, depending on how you rescale the graph in your lab report or presentation.

- Professional plots do NOT have titles.

Resultant Plot:

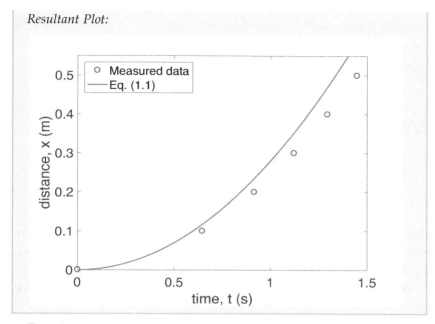

Exercises 1:

1. Describe Eratosthenes work to determine the radius of earth in terms of the 6-step scientific method. What was the question he sought to answer? What information did he find in his reading? What was his hypothesis? What was the experiment he performed? How did he process the results? How do we even know that he performed this experiment 2000 years ago?

2. An interesting riddle told at engineering parties involves the "waddle" or "dewlap" of a male moose. People are asked to explain why male moose have a large mass of tissue hanging from their chin *without looking it up on the Internet.* Formulate a hypothesis for why bull moose have dewlaps, and propose an experimental observation that could be used to test your hypothesis.

Dewlap

3. Shown to the left is a table containing measured values of initial velocity v_0 and maximum height h attained by a projectile fired vertically into the air. Using the format outline in Example 1.1, plot the measured height as a function of the measured initial velocity (i.e. height on the vertical y-axis and initial velocity on the horizontal x-axis). Use conservation of energy to derive a theoretical equation and add that to the plot that as a solid continuous line. Export the plot from Matlab as a PDF, BMP, or EPS file, import it into LaTeX or Microsoft Word and put a descriptive caption below the plot.

v_0 (m)	h (m)
0	0
1	0.05
2	0.201
3	0.452
4	0.803
5	1.255
6	1.807
7	2.460
8	3.212

2 Measurements and Transducers

2.1 SI Units

The **international system of units (SI units)** is widely used by scientists and engineers around the world to quantify various physical phenomena. Importantly, SI units have the advantage that they are defined in such a way that certain physical constants work out to be nice, even numbers.

- The density of water $\rho_w = 1 \text{ g/cm}^3 = 1000 \text{ kg/m}^3$.

- The period of a 1 meter long pendulum is $T = 2$ s.

- Water melts at $0°$ Celsius and boils at $100°$ Celsius.

This book will primarily use SI units. However, we will sometimes use imperial units of pounds per square inch (psi) or inches of water (in. H_2O) for pressure. Machine shops in America typically prefer to work in inches, so it is common for CAD drawings to be dimensioned in inches.

2.2 Measurements and Transducers

Science and engineering require us to measure and quantify a variety of abstract phenomena, including time, temperature, speed, acceleration, force, pressure, stress, and strain, just to name a few. As human beings, our brains are very good at processing spatial information. This makes it easy for us to think about *distance* and quantify it. However, other phenomena—such as pressure or air speed—are more abstract and difficult to mentally quantify.

A **transducer** is used to convert an abstract physical phenomenon into a visually observable distance. For example, an alcohol thermometer converts temperature to a distance. The alcohol in the thermometer expands or contracts with temperature, causing it to rise or descend to a certain length in a glass tube.

2.3 Manometer

The **manometer** is a transducer that converts pressure to length. Our
eyes allow us to easily see distance and quantify it. Our ears allow us
to sense pressure, but not in a very quantifiable way. Anyone who has
swam to the bottom of a deep pool likely felt a slight pain in their ears.
This pain is due to the **hydrostatic pressure** of the water above.

Hydrostatic pressure is pressure due to the weight of a fluid. As the
fluid gets deeper, the hydrostatic pressure increases

$$p = \rho_L g h, \tag{2.1}$$

Notes: Hydrostatic pressure is
often referred to as "pressure
head".

where ρ_L is the density of the fluid, $g = 9.8$ m/s^2 is the acceleration
due to gravity, and h is the depth of the fluid.

Equation (2.1) shows us that hydrostatic pressure is directly proportional to a distance h. The U-tube manometer shown in Fig. 2.1
exploits this to create a transducer. When the pressure difference
$\Delta p = p_1 - p_2 = 0$, there is no difference in the liquid height. When
$p_1 > p_2$, the liquid is pushed down in the left side and pushed up on
the right side.

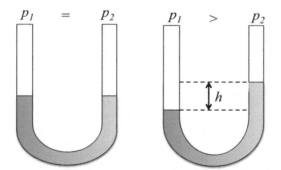

Figure 2.1: A U-tube manometer contains a blue liquid. The difference in liquid height
is proportional to the difference in pressure.

Notes: Pressure is always measured as a difference between
two points, $\Delta p = p_1 - p_2$.

- Pressure relative to the vacuum of space is called **absolute pressure**.

- Pressure relative to the surrounding atmospheric air
  is called **gauge pressure**.

The measured height h can be plugged into Eq. (2.1) to calculate
the pressure in psi or Pascal. However, it is common to simply leave
the pressure in units of distance, such as "inches of water" or "mm of
mercury", depending on the fluid.

2.4   Sensitivity and Range

Sensitivity and range are important properties of any transducer.

- **Sensitivity** is the change in length divided by the change in the measured parameter. For the manometer, it is $\frac{dh}{dp} = \frac{1}{\rho_L g}$.

- **Range** is the maximum and minimum values that a transducer can measure. For a manometer, the range is $h_{max}\rho_L g$, where h_{max} is the maximum vertical height of the tube.

There is always a trade-off between sensitivity and range. Increasing the range typically decreases the sensitivity, while increasing the sensitivity typically decreases the range. For example, a manometer filled with mercury (Hg) will have more range, but less sensitivity than a manometer filled with water.

2.5   Pitot Tube

The **pitot tube** or **pitot static probe** is a transducer that converts airspeed u to pressure using the Bernoulli effect

$$\Delta p = \frac{1}{2}\rho_A u^2, \tag{2.2}$$

where $\rho_A = 1.23$ kg/m^3 is the density of air at room temperature at sea level.

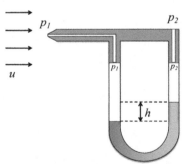

Figure 2.2: A gray pitot tube (top) is connected to a U-tube manometer containing a blue liquid (bottom).

Connecting the pitot tube to a manometer, as shown in Fig. 2.2, creates a transducer that converts airspeed to a measurable length. Combining Eqs. (2.1) and (2.2) and solving for the airspeed u yields

$$u = \sqrt{\frac{2\rho_L g h}{\rho_A}}. \tag{2.3}$$

Note that the pitot tube has two ports. One must be facing directly into the moving air to measure the **stagnation pressure** p_1. The other must be perpendicular to the moving air to measure the **static pressure** p_2.

Example 1.1:

A pitot static probe connected to a manometer is used to measure air speed in a wind tunnel. What is the airspeed in m/s if the manometer reads $h = 0.8$ inches of water?

Solution: First, convert to SI units: $h = 0.8$ inches $= 0.020$ m

Then, use Eq. (2.3)

$$u = \sqrt{\frac{2(1000kg/m^3)(9.8m/s^2)(0.02m)}{1.23kg/m^3}} = 17.9 \text{ m/s}$$

Exercises 2:

1. Convert the following pressures to SI units of Pascal.

   (a) 3 inches of water

   (b) 30 inches of mercury

   (c) 10 psi

   (d) 1 atmosphere

   (e) 2 bar

   (f) 100 Torr

2. A U-tube manometer is filled with water and has a maximum height of 9 inches of water.

   (a) Calculate its sensitivity (in meter/Pascal) and range (in Pascal).

   (b) Consider the same manometer filled with mercury instead of water. Calculate the new sensitivity and range.

3. A pitot probe connected to a manometer is used to measure airspeed. At a certain airspeed, the manometer reads 1 inch of water. If the airspeed is doubled, how many inches of water will the manometer read?

3 Prelude to Electronics

As we discussed last chapter, the *human brain* is very good at processing spatial information, and many traditional transducers convert the measured phenomenon to a length. *Computers* are very good at processing electrical signals—voltages in particular. Electronic sensors that output a voltage are popular, because they can be directly interfaced with a computer system.

An **analog transducer** or **analog sensor** is a device that converts a physical phenomenon into an observable voltage.

3.1   Review of Electricity and Magnetism

To properly understand and utilize electronic sensors, it is important to understand the physical principles that allow them to work. We will begin with a brief review of electricity and magnetism. (It is assumed that the reader has taken or is currently enrolled in a Physics II course that covers basic electricity and magnetism.)

Electric Charge q is a fundamental physical property of any particle.

- Charge can be positive or negative.
- Opposite charges are attracted, like charges repel one another.
- The SI unit for charge is the Coulomb [C].
- The electron has a charge of -1.6×10^{-19} C, while the proton has a charge of $+1.6 \times 10^{-19}$ C.

Electric Potential ϕ is the electrical potential energy. Shown in Fig. 3.1, positive charge moves from high potential to low potential (similar to how the ball moved from high to low gravitational potential in Galileo's inclined plane).

- Electric potential has units of energy per charge.
- The SI unit for electric potential is the Volt [V] = [1 J/C].

Voltage V is the *difference* in electric potential energy at two different locations in space. Shown in Fig. 3.1, it is illustrated as $V = \phi_1 - \phi_2$.

- Voltage is measured with a **voltmeter** or an *analog-to-digital converter (A/D)* .

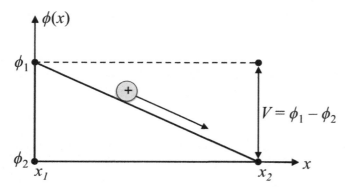

Figure 3.1: Positive charge is accelerated from high to low electric potential.

Electric Field $E = -\frac{d\phi}{dx}$ is the negative spatial gradient of electric potential. The force on a charged particle due to and electric field is $F = qE$.

Current $I = \frac{dq}{dt}$ is the rate that charge moves through a medium, typically a wire.

- Current can be positive or negative depending on the direction that the charge is moving.

- The SI unit for current is the Ampere or "Amp" [A] = [1 C/s].

- Current is measured with an **ammeter**.

A **Magnetic Field** B can be produced by moving charge (current) or by a changing electric field. Charged particles only respond to a magnetic field if they are moving. Thus coils of wire that carry current are subject to the magnetic force, which is how electric motors work.

- The SI unit for magnetic field is the Tesla [1 T] = [1 Vs/m^2] = [1 kg/As^2].

- A Hall sensor can be used to measure magnetic field strength.

- The torque in an electric motor comes from a magnetic force.

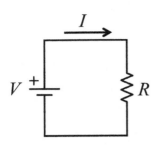

Figure 3.2: A resistor (right) is connected to a DC power supply (left).

3.2   Ohm's Law

A **direct current (DC) power supply** is a device that produces a constant electric potential difference—or voltage—between two output terminals. A DC power supply is connected to a **resistor** in the circuit shown in Fig. 3.2, and the voltage causes current to flow through the resistor. The amount of current flowing out of the power supply and through the resistor is related to the voltage drop across the resistor by **Ohm's Law**

$$V = IR, \tag{3.1}$$

where R is the resistance of the resistor. Resistance has units of Ohms $[\Omega] = [1 \text{ V}/\text{A}]$, and it can be measured using an **ohmmeter** .

3.3   Joule Heating

A voltage difference causes electrons to move through a material. As they move through the material, they collide with atoms and transfer kinetic energy, as illustrated in Fig. 3.3. This transfer of kinetic energy is known as **Joule Heating** . This will cause the temperature of the resistor to increase.

Notes:

- Capital Q is used to denote thermal energy or "heat".
- Lower case q usually denotes charge.

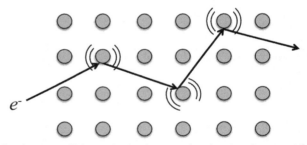

Figure 3.3: An electron collides with metal atoms, thus heating the material.

Notes: Joule heating is usually considered an undesirable energy loss in many consumer electronics devices.

The rate of heating, or "power dissipated", can be calculated by multiplying current times voltage

$$\dot{Q} = IV. \tag{3.2}$$

Combing Eqs.(3.2) and (3.1) gives the power dissipated as a function of resistance

$$\dot{Q} = I^2 R, \tag{3.3}$$

where I is the current through the resistor, or

$$\dot{Q} = \frac{V^2}{R}. \tag{3.4}$$

where V is the voltage dropped across the resistor. Any of these three formulas are valid for calculating the power dissipated in a resistor.

Example 3.1:
Consider the circuit shown in Fig. 3.1. Calculate the power dissipated in the resistor if $V = 12V$, $I = 0.12A$, and $R = 100\Omega$.

Solution:
Using Eq. (3.2): $\dot{Q} = (0.12A)(12V) = 1.44W$

Using Eq. (3.3): $\dot{Q} = (0.12A)^2(100\Omega) = 1.44W$

Using Eq. (3.4): $\dot{Q} = (12V)^2/(100\Omega) = 1.44W$

Success! All three equations give the same result.

3.4   Digital Multimeter (DMM)

The **digital multimeter (DMM)** is a device that can be used to measure current, voltage, and resistance. Illustrated in Fig. 3.4, the DMM can function as an ammeter to measure current, a voltmeter to measure voltage, or an Ohmmeter to measure resistance.

Notes:

- The ammeter is always connected in series, so current may flow through it.

- The voltmeter is always connected in parallel, and the probes measure the voltage *difference* between two points in the circuit.

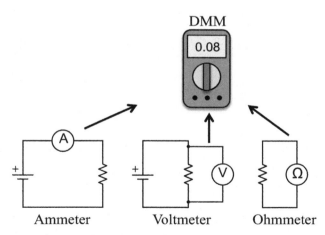

Figure 3.4: A DMM can measure current (left), voltage (center), and resistance (right).

Exercises 3:

1. Refer to Fig. 3.1. Calculate the electric field in units of V/m if $x_1 = 1.2cm$, $x_2 = 1.9cm$, $\phi_1 = 30V$, and $\phi_2 = 6V$.

2. Show that Eq. (3.2) gives the power dissipated in units of Watts [J/s] if current in is Amps and voltage is in Volts.

3. Refer to the circuit shown in Fig. 3.2. Assume $V = 5V$ and $R = 220\Omega$.

   (a) Calculate the current I.

   (b) Calculate the power dissipated in the resistor.

4 DC Circuits

Electronic sensors and transducers are always connected to some electronic circuit. In this lecture, we will review basic **DC circuits** .

4.1 Kirchhoff's Circuit Laws

When designing or constructing an electronic circuit, it is important to know the electrical current and voltage at various locations in the circuit. These parameters can be calculated using **Kirchhoff's Circuit Laws.**

1. The **current law** states that the current flowing into a junction is equal to the current flowing out of the junction. Shown in Fig. 4.1, we have the relationship $I_1 + I_2 = I_3 + I_4$ for the currents flowing into the junction or "node".

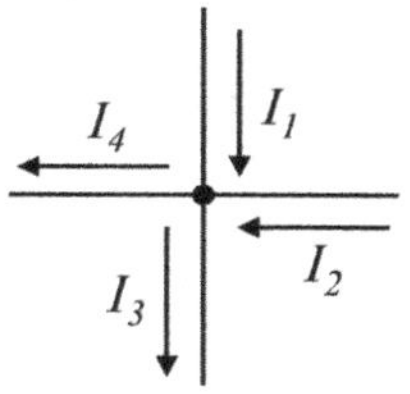

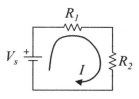

Figure 4.1: Current flows into and out of a junction.

2. The **voltage law** states that the sum of voltages around a closed loop is zero. For the circuit shown in Fig. 4.2, we assume the current flow clockwise around the loop. Tracing the current flow around the loop, we add up the voltage drops. The resultant equation is $V_s - IR_1 - IR_2 = 0$.

   - The direction of the loop goes from the short to the long horizontal bar on the DC power supply, so we add positive V_s.

   - The direction of the loop goes in the same direction of the current flowing through the resistors, so we add negative IR_1 and IR_2.

Figure 4.2: Current flows clockwise around the "loop".

It is worth noting that the current law is essentially the law of conservation of charge, and the voltage law represents conservation of energy.

Example 4.1:

Use Kirchhoff's Circuit Laws to calculate the four currents I_1, I_2, I_3, and I_4 shown in the circuit below, where $V = 5\text{V}$, $R_1 = 60\Omega$, $R_2 = 80\Omega$, $R_3 = 40\Omega$, and $R_4 = 40\Omega$.

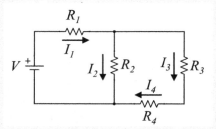

Solution: We will use Kirchhoff's Laws to obtain four linearly independent equations, and solve them for the four unknowns I_1, I_2, I_3, and I_4.

Applying the Current Law to the top node, we obtain
$I_1 = I_2 + I_3$.
Applying the Current Law to the bottom node, we obtain
$I_1 = I_2 + I_4$.
Applying the Voltage Law to the left loop, we obtain
$V - I_1 R_1 - I_2 R_2 = 0$.
Applying the Voltage Law to the right loop, we obtain
$I_2 R_2 - I_3 R_3 - I_4 R_4 = 0$.

Voila! We now have four equations and four unknowns. To solve the system, we will first write it in matrix form.

$$\begin{bmatrix} 0 \\ 0 \\ V \\ 0 \end{bmatrix} = \begin{bmatrix} 1 & -1 & -1 & 0 \\ 1 & -1 & 0 & -1 \\ R_1 & R_2 & 0 & 0 \\ 0 & R_2 & -R_3 & -R_4 \end{bmatrix} \begin{bmatrix} I_1 \\ I_2 \\ I_3 \\ I_4 \end{bmatrix}$$

Note that this is essentially a 4-dimensional version of Ohm's law $\vec{V} = \mathbf{R}\vec{I}$, where $\mathbf{R}$ is a 4×4 matrix, and $\vec{V}$ and $\vec{I}$ are column vectors.

Using Matlab, we can easily invert the matrix $\mathbf{R}$ to determine the currents, $\vec{I} = \mathbf{R}^{-1}\vec{V}$.

Matlab Script:

```
 1 -    clc
 2 -    close all
 3
 4 -    Vs=5;     %volts
 5 -    R1=60;    %Ohms
 6 -    R2=80;    %Ohms
 7 -    R3=40;    %Ohms
 8 -    R4=40;    %Ohms
 9
10 -    R=[1 -1 -1 0;
11          1 -1 0 -1;
12          R1 R2 0 0;
13          0 R2 -R3 -R4]  %R matrix
14
15 -    V=[0; 0; 5; 0]  %V column vector
16
17 -    I=R^-1*V     %Output solution
```

Matlab Output:

```
I =

    5.0000e-02
    2.5000e-02
    2.5000e-02
    2.5000e-02
```

Thus, we have $I_1 = 50$mA, $I_2 = I_3 = I_4 = 25$mA.

4.2   Voltage Divider Circuit

As we will see Lecture 6, many transducers convert a measurable physical parameter to an electrical resistance. This resistance can then be converted to a voltage using the **voltage divider circuit** shown in Fig. 4.3.

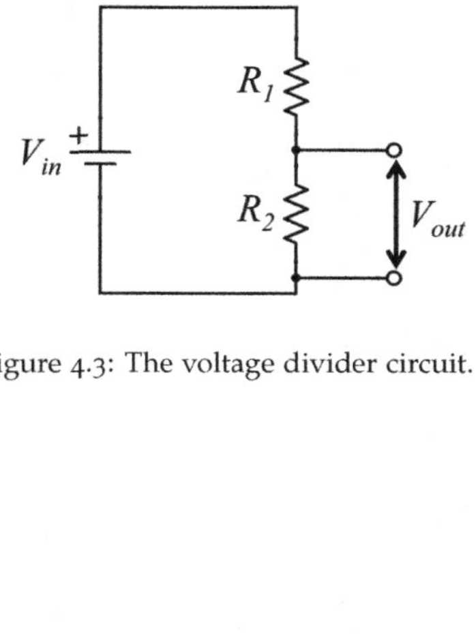

The output voltage of the circuit V_{out} is related to the input voltage V_{in} and the resistances R_1 and R_2 via

Figure 4.3: The voltage divider circuit.

$$V_{out} = \frac{R_2}{R_1 + R_2} V_{in}. \tag{4.1}$$

We will revisit the voltage divider circuit again in Lecture 6, where one of the resistors will be replaced with a resistive transducer.

4.3   Wheatstone Bridge Circuit

Another circuit that is useful for resistive transducers—strain gauges in particular—is the **Wheatstone Bridge Circuit** shown in Fig. 4.4.

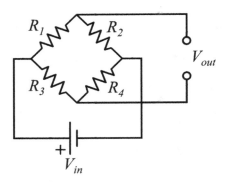

Note:

- Resistors in series can be added to obtain an equivalent resistance $R_{eq} = R_1 + R_2$.

- For resistors in parallel, we add the inverses $R_{eq}^{-1} = R_1^{-1} + R_2^{-1}$.

Figure 4.4: The Wheatstone bridge circuit.

The output voltage of the circuit V_{out} is related to the input voltage V_{in} and the resistances R_1, R_2, R_3, and R_4 via

$$V_{out} = \left(\frac{R_2}{R_1 + R_2} - \frac{R_4}{R_3 + R_4} \right) V_{in}. \tag{4.2}$$

Note that the output $V_{out} = 0$ if $R_1/R_2 = R_3/R_4$, and we say that the bridge is "balanced".

In Lecture 6, we will see that the Wheatstone bridge circuit is particularly good for resistive transducers known as strain gauges.

4.4   Grounded Circuits

As noted in Lecture 3, voltage is the difference between two electrostatic potentials, $V = \phi_1 - \phi_2$. It is a standard convention to assume that the electrostatic potential of our planet Earth is zero. This makes measurements more simple, as $\phi_1 = \phi_{earth} = 0$. Additionally, humans typically stand on the ground, so we are also at zero volts, which makes it easier to design safe electrical circuits.

Thus, many circuits are **grounded circuits!grounded**, where there is literally a wire going into the ground that either pulls charge out of the earth or dumps charge back into the earth, as illustrated by the current loop on the left of Fig. 4.5.

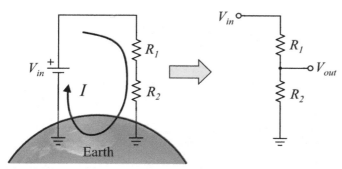

Figure 4.5: A voltage divider circuit is referenced to ground. It is assumed that the electrostatic potential of Planet Earth is zero.

The voltage divider shown in Fig. 4.3 is what we would call a "floating" circuit. However, the same circuit could be grounded as shown on the right in Fig. 4.5. (The three horizontal bars that form an inverted pyramid represent a connection to earth ground.) To construct the grounded circuit, one would need to connect the negative terminal of the power supply to ground. The DMM used to measure V_{out} would also need to have its negative terminal connected to ground.

Exercises 4:

1. In the circuit shown below $V = 12V$, $R_1 = 220\Omega$, $R_2 = 470\Omega$, $R_3 = 470\Omega$, and $R_4 = 100\Omega$.

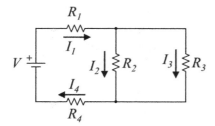

Calculate the following.

   (a) The current I_1.

   (b) The current I_3

   (c) The power dissipated in R_4

2. Use Kirchhoff's circuit Laws to derive the voltage divider equation, Eq. (4.1).

3. Use algebra to show that the output of the Wheatstone bridge $V_{out} = 0$ if $R_1/R_2 = R_3/R_4$.

5 DC Power Supplies

5.1 Non-Ideal Power Supplies

In Lectures 3 and 4, we only considered *ideal* DC power supplies that can output any amount of current needed by the circuit. However, this is not true in the real world.

In reality, there is a limit on the amount of current that a power supply can produce. To model this, we assume that a DC power supply has an **internal resistance** R_S that limits the amount of current it can produce. Shown in the dashed lines in Fig. 5.1, the **non-ideal power supply** is modeled as an ideal power supply with voltage V_{OC} in series with a resistor R_S, and it is used to power a "resistive load" R_L.

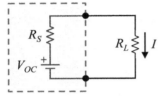

Figure 5.1: A non-ideal DC power supply (left) drives a resistive load R_L (right).

Notes: A "resistive load" is a placeholder for anything that needs electrical power. It could be a light bulb, sensor, motor, audio speaker, etc.

5.2 Short Circuit Current

A non-ideal power-supply has a limit on the amount of current it can output. If the load resistance $R_L = 0$, the amount of current will be limited by the internal resistance R_S of the power supply. This current is known as the **short circuit current** $I_{SC} = V_{OC}/R_S$.

Short Circuit: $R_L = 0$

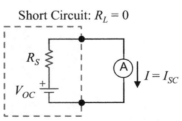

Figure 5.2: The power supply is short circuited through an ammeter. It is assumed that the ammeter has a resistance $R_L = 0$.

Notes: Short circuits are very dangerous. The high current heats up the wires, which can cause a fire.

Shown in Fig. 5.2, the load resistor is replaced with an ammeter, which has a resistance of zero. The resultant current flowing through the ammeter is the short circuit current I_{SC}.

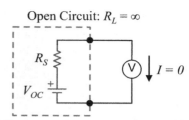

Open Circuit: $R_L = \infty$

Figure 5.3: The open circuit voltage is measured with a voltmeter. It is assumed that the voltmeter has an infinite resistance $R_L = \infty$.

5.3   Open Circuit Voltage

If the load resistance $R_L = \infty$, then the voltage output measured across the two output terminals is the **open circuit voltage**. Shown in Fig. 5.3, the load resistor is replaced with a voltmeter, which has infinite resistance. The measured voltage is V_{OC}.

5.4   Voltage Droop

Applying the voltage law to the circuit shown in Fig. 5.1, we obtain

$$V_{OC} - IR_S - IR_L = 0. \tag{5.1}$$

Solving for the current I yields

$$I = \frac{V_{OC}}{R_S + R_L}. \tag{5.2}$$

Notes: Internal resistance is often referred to as the "output impedance" of the power supply.

Note that for $R_L = 0$, Eq. (5.2) gives us the shorts circuit current $I_{SC} = V_{OC}/R_S$. The voltage drop across the load $V_L = IR_L$ then becomes

$$V_L = \frac{R_L}{R_S + R_L} V_{OC}. \tag{5.3}$$

The load voltage V_L is the effective voltage output by the non-ideal power supply. Interestingly, it depends on the load resistance R_L.

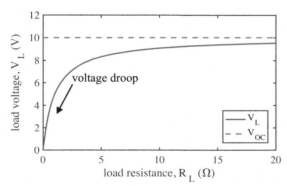

Figure 5.4: Equation (5.3) is plotted as a function of R_L for $R_S = 1\Omega$ and $V_{OC} = 10V$.

Equation (5.3) is plotted in Fig. 5.4 for a 10V DC power supply with $R_S = 1\Omega$, and we see that the load voltage actually decreases for lower load resistances. This is a phenomenon known as **voltage droop** —the output voltage of a power supply will decrease when it outputs too much current.

5.5 Impedance Matching

As engineers, we harness energy to perform work, so we are always interested in efficient power transmission. The power transmitted from the DC power supply to the resistive load R_L can be calculated by multiplying Eqs. (5.2) and (5.3) together, $\dot{Q}_L = IV_L$, or

$$\dot{Q}_L = \frac{R_L V_{OC}^2}{(R_S + R_L)^2}. \tag{5.4}$$

Similar to the voltage output, we see that the power output depends on the load resistance R_L. Equation (5.4) is plotted in Fig. 5.5 for a 10V DC power supply with $R_S = 1\Omega$, and we see that the output power has a maximum value when the load resistance *matches* the internal resistance, $R_L = R_S = 1\Omega$. This is a phenomenon known as **impedance matching**.

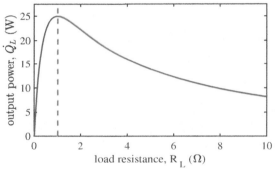

Figure 5.5: Equation (5.4) is plotted as a function of R_L for $R_S = 1\Omega$ and $V_{OC} = 10V$.

Example 5.1:

A solar panel is being tested by an engineer. She measures short circuit current $I_{SC} = 0.7A$ and open circuit voltage $V_{OC} = 21V$

a) What is the internal resistance?

Solution: $R_S = V_{OC}/I_{SC} = 21V/0.7A = 30\Omega$

b) The solar panel is used to power an electronic device with resistance R_L. What should the resistance R_L be to maximize the power output?

Solution: $R_L = R_S = 30\Omega$

c) What is the maximum power that the solar panel can output?

Solution: $\dot{Q}_L = \frac{R_L V_{OC}^2}{(R_S + R_L)^2} = \frac{30\Omega}{(60\Omega)^2}(21V)^2 = 3.68W.$

Exercises 5:

1. A 9 V battery has an internal resistance of 10Ω, and it is used to power a small light bulb.

   (a) What should the resistance of the light bulb be in order to maximize its brightness?

   (b) What is the maximum power that can be delivered to the light bulb?

2. Use calculus to show that Eq. (5.4) has a maximum at $R_L = R_S$. (Hint: Begin by taking the derivative and setting it equal to zero, $\frac{d\dot{Q}_L}{dR_L} = 0$).

3. Physicist Geoffrey West and his colleagues have developed a famous theory that electricity and other resources are distributed most efficiently through *impedance matched fractal networks*. Common examples of this are electrical grids, cardiovascular systems in animals, roots and branches of plants, HVAC systems in large buildings, and pipes for distributing water and natural gas.

   (a) Consider the network below, which contains $N = 3$ levels of branches. Each individual resistor in the network has a resistance R. Derive an equation for the equivalent resistance R_{eq} of the entire network in terms of R.

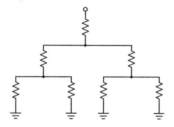

   (b) Based on your answer from part a, derive an equation for the equivalent resistance of a similar network with an arbitrary number of levels N, where all resistors have the same resistance R.

   (c) What happens to the equivalent resistance as N goes to infinity?

6 Resistive Sensors

In this lecture, we will introduce several transducers that convert a physical parameter to an electrical resistance. These resistances are typically converted to voltages using an electrical circuit such as the voltage divider or Wheatstone bridge seen previously in Lecture 4.

6.1 RTD

The **resistance temperature detector (RTD)** is a thin metal strip of wire (usually platinum) encased in plastic epoxy. The resistance of the metal strip R increases linearly with temperature T. Thus, by measuring the resistance of the RTD, one can then calculate the temperature using the formula

$$T = T_0 + \left(\frac{R - R_0}{\alpha_T R_0} \right), \tag{6.1}$$

Notes: Platinum has a TCR of 0.00385/°C or 3850ppm/°C.

where α_T is the temperature coefficient of resistance (TCR) and R_0 is the resistance at some reference temperature T_0. (Typically, the reference temperature $T_0 = 0°$ C.)

Example 6.1:
An RTD is wired up in a voltage divider circuit, where R_s is the RTD, $V_{in} = 5$V, and $R_f = 220\Omega$.

Notes: In a circuit diagram, a resistive sensor is typically denoted as a resistor with an arrow going through it.

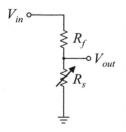

According the manufacturer's data sheet, the RTD has parameters $\alpha_T = 0.00385/°C$, $R_0 = 100\Omega$, and $T_0 = 0°C$. Calculate the measured temperature T, if the output voltage $V_{out} = 1.9V$.

Solution: First, use the voltage divider equation to solve for the resistance R_s of the RTD in terms of the given V_{out}.

Voltage divider equation: $V_{out} = \frac{R_s}{R_s+R_f}V_{in}$.

Solving for R_s yields $R_s = \frac{V_{out}R_f}{V_{in}-V_{out}} = \frac{(1.9V)(220\Omega)}{3.1V} = 134.8\Omega$

Now, use Eq. (6.1) to calculate the temperature.

$$T = T_0 + \left(\frac{R-R_0}{\alpha_T R_0}\right) = 0°C + \left(\frac{134.8\Omega-100\Omega}{(0.00385/°C)(100\Omega)}\right) = 90.4°C$$

6.2 Thermistor

The **thermistor** is like the RTD, except it is made of a semiconductor or a ceramic rather than a metal. The resistance of the semiconductor or ceramic R increases nonlinearly with temperature T. By measuring the resistance of the thermistor, one can then calculate the temperature using the Steinhart formula

Notes: Negative temperature coefficient (NTC) means the thermistor's resistance will *decrease* with temperature.

$$T = \left[A + B\ln\left(\frac{R}{R_{ref}}\right) + C\ln\left(\frac{R}{R_{ref}}\right)^2\right]^{-1}, \qquad (6.2)$$

where A, B, C, and R_{ref} are calibration constants that can typically be found in the manufacturer's data sheet.

6.3 Strain Gage

The electrical resistance of a strip of metal wire is given by the formula

$$R = \rho\frac{L}{A}, \qquad (6.3)$$

where ρ is the resistivity of the material, L is the length of the wire, and A is the cross sectional area of the wire.

Mechanical strain is defined as the change in length divided by the original length of a material, $\epsilon = \Delta L/L$. If a wire is stretched under tension, its electrical resistance will increase. If the wire is compressed, its electrical resistance will decrease. This is the physical basis of the **strain gage**. A long thin strip of wire is encased in a polymer film and glued to the surface of a mechanical part. As the surface is stretched

Notes: Strain gages are glued to metal surfaces using "cyanoacrylate glue", a.k.a. superglue.

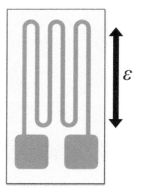

Figure 6.1: An illustration of a strain gage.

Notes: Strain is a tensor that is defined by the orientation of the surface and the direction of the stretching or compression.

or compressed, the resistance of the strain gage changes according to the formula

$$\frac{\Delta R}{R} = G_f \frac{\Delta L}{L} = G_f \epsilon, \qquad (6.4)$$

where R is the resistance of the strain gage under zero strain and G_f is the gage factor. The gage factor is a calibration constant that can typically be found in the manufacturer's data sheet, and it typically has a value of approximately 2.

Illustrated in Fig. 6.1, the metal wire zigzags back and fourth several times through the plastic film to increase its sensitivity. The strain gage measures strain in the direction of the long thin strips of wire. The two large squares at the ends of the wire are called "solder pads", and wires are soldered to these so the strain gage may be connected to a circuit.

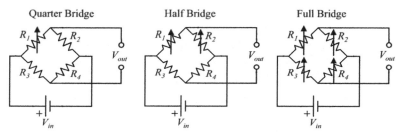

Figure 6.2: Strain gages can be wired up in various different Wheatstone bridge configurations. Arrows denote which resistors have been replaced with strain gages.

Strain gages are often connected in a Wheatstone bridge circuit, as shown in Fig. 6.2. There are three different configurations involving different numbers of strain gages.

- A **quarter bridge** has only one of the resistors replaced with a strain gage.

- A **half bridge** has two of the resistors replaced with strain gages.

- A **full bridge** has all of the resistors replaced with a strain gages.

Equations (4.2) and (6.4) can be used to calculate the strain from the measured output voltage for the various bridge configurations. Alternatively, one can purchase a commercial "strain gage amplifier", such as Vishay's P3, to automatically balance the bridge and calculate the strain.

Example 6.2:

Two strain gages are mounted to a cantilever beam, as shown below, and connected in the half bridge circuit shown in Fig. 6.2. The top strain gage is wired up as R_2 and the bottom strain gage is wired up as R_1. Both strain gages have a resistance $R = 120\Omega$ when there is zero strain. The other two resistors in the half bridge circuit are chosen to have values $R_3 = R_4 = R = 120\Omega$.

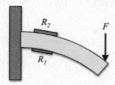

Use Eqs. (4.2) and (6.4) to show that the strain is related to the output voltage by the equation $\epsilon = \frac{2V_{out}}{G_f V_{in}}$.

Solution:

The top strain gage R_2 is stretched, so it resistance *increases* by an amount ΔR:

$R_2 = R + \Delta R$.

The bottom strain gage R_1 is compressed, so it resistance *decreases* by an amount ΔR:

$R_1 = R - \Delta R$.

The change in resistance ΔR will be related to the strain ϵ given by Eq. (6.4).

Substituting the above equations into the Wheatstone bridge equation, Eq. (4.2), we obtain $\frac{V_{out}}{V_{in}} = \left(\frac{R+\Delta R}{R-\Delta R+R+\Delta R} - \frac{R}{2R} \right)$

Cancelling terms in the denominator and adding the fractions, we obtain the remarkably simple equation $\frac{V_{out}}{V_{in}} = \frac{\Delta R}{2R}$.

Then, substituting Eq. (6.4) gives us $\frac{V_{out}}{V_{in}} = \frac{G_f \epsilon}{2}$.

Lastly, solving for ϵ yields $\boxed{\epsilon = \dfrac{2V_{out}}{G_f V_{in}}}$

6.4   Potentiometers

The **potentiometer** is a transducer that converts the angular rotation of a shaft to an electrical resistance. Illustrated on the left of Fig. 6.3, it consists of an arc of resistive material that makes contact with a brush that is mechanically coupled to a rotating shaft. Rotating the shaft moves the brush along the arc, which changes the arc length and the resistance between A and B and between B and C.

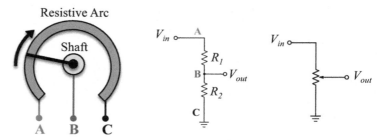

Figure 6.3: An illustration of a potentiometer and its use as a circuit element.

The potentiometer can be used as a single variable resistor by connecting only the A and B terminals. Rotating the shaft clockwise increases the resistance between A and B. The potentiometer can also be used as a variable voltage divider, as illustrated in the center of Fig. 6.3. In this configuration, R_1 is the resistance between A and B, and R_2 is the resistance between B and C. Rotating the potentiometer clockwise increases R_1 and simultaneously decreases R_2. This effectively changes the output voltage V_{out}. The schematic symbol for the potentiometer is shown on the right of Fig. 6.3.

6.5   Photoresistor

Notes: The sensitivity of a photoresistor depends on the color, or wavelength, of the light. Take care to select one that matches the color of the measured light.

Photoresistors or "photocells" are sensors made of a photosensitive material whose resistance depends on light intensity. A commonly used photoresistive material is cadmium sulfide (CdS). These sensors show up in many consumer electronic devices. In particular, they are found in electric lights that automatically turn on when the sun goes down. (This is an excellent example of how a sensor can be used to automate a simple process.)

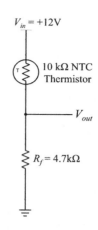

$V_{in} = +12V$

10 kΩ NTC Thermistor

V_{out}

$R_f = 4.7k\Omega$

Exercises 6:

1. A thermistor is wired up in a voltage divider, as shown to the left. The manufacturer's data sheet gives the calibration parameters $R_{ref} = 10k\Omega$, $A = 3.354 \times 10^{-3}$ K^{-1}, $B = 2.57 \times 10^{-4}$ K^{-1}, and $C = 2.62 \times 10^{-6}$ K^{-1}. Use the voltage divider equation and Steinhart equation to calculate the measured temperature T if $V_{out} = 9.4V$.

2. Consider the RTD in the voltage divider circuit from Example 6.1. Redesign the circuit so it outputs 3V at 100°C. That is, calculate the fixed resistance R_f that would achieve this.

3. Consider the cantilever beam pictured to the left. The beam has four identical strain gages, each with a resistance $R = 120\Omega$ when no strain is applied. The strain gages are wired up in a full bridge conifuration with R_2 and R_3 on top and R_3 and R_4 on bottom. Use Eqs. (4.2) and (6.4) to show that the strain is related to the output voltage by the equation $\epsilon = \frac{V_{out}}{G_f V_{in}}$.

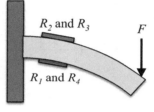

R_2 and R_3

F

R_1 and R_4

7 *Capacitive Sensors*

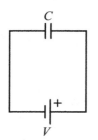

Figure 7.1: A capacitor is charged by a DC power supply.

Capacitive transducers convert a physical parameter to an electrical capacitance. The capacitance is then converted to an analog voltage using fairly sophisticated circuits. These tertiary circuits are beyond the scope of this book, so we will primarily focus on the physical mechanisms that change the capacitance.

7.1 *Capacitors*

Capacitors store charge and electrical energy. Shown in Fig. 7.1, a capacitor with capacitance C is connected to a DC power supply with voltage V, causing the capacitor to become charged. The amount of charge q on the capacitor is given by

$$q = CV. \tag{7.1}$$

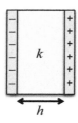

Figure 7.2: An illustration of a parallel plate capacitor.

Charging a capacitor requires work from the DC power supply. This work is stored as potential energy. The electrical energy stored in a capacitor is given by

$$U = \frac{q^2}{2C}, \tag{7.2}$$

or, substituting Eq. (7.1) for q,

$$U = \frac{1}{2}CV^2. \tag{7.3}$$

The simplest geometry for a capacitor is the **parallel plate capacitor**, which is illustrated in Fig. 7.2. The capacitance of a parallel plate capacitor is given by

$$C = \frac{k\epsilon_0 A}{h}, \tag{7.4}$$

where k is the dielectric constant, ϵ_0 is the permittivity of free space, A is the surface are of the plates, and h is the distance between the plates.

7.2   Capacitive Pressure Transducer

The **capacitive pressure transducer** is a parallel plate capacitor, where at least one of the electrode plates is made of some elastic material. Illustrated in Fig. 7.3, applying a differential pressure $p_1 > p_2$ causes the elastic plate to deflect inward, thus reducing the gap distance by an amount Δh and increasing the capacitance by an amount ΔC. An external circuit would then convert this capacitance to a voltage, resulting in a voltage output that is linearly related to the pressure difference $\Delta p = p_1 - p_2$.

Capacitive pressure transducers can be purchased in a variety of different configurations:

- **Differential pressure** means it has two ports for p_1 and p_2.

- **Vented gauge** means it only has one port for p_1, and p_2 is simply vented to the ambient air.

- **Absolute** means the space between the plates is a vacuum, such that $p_2 = 0$.

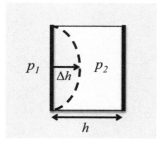

Figure 7.3: Applying a pressure differential $p_1 > p_2$ causes the left electrode to deform, decreasing the gap distance h and increasing the capacitance C.

7.3   MEMS Accelerometer

Micro-electromechanical systems (MEMS) devices are tiny mechanical structures made from etched silicon and metal. They are often smaller than the width of a human hair. Illustrated in Fig. 7.4, a MEMS accelerometer usually consists of interdigitated electrodes that form a capacitor. One of the electrodes has a mass m and is attached to a spring with spring constant k. When the device is accelerated, the inertia of the electrode pushes on the spring with a force $F = ma$, causing the electrode to be displaced by an amount Δh and the capacitance to change by an amount ΔC.

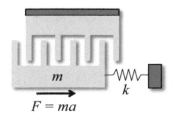

Figure 7.4: The MEMS accelerometer has interdigitated electrodes.

Exercises 7:

1. A capacitor is charged by a DC power supply, as shown in Fig. 7.1. Calculate the following if $C = 100\text{nF}$ and $V = 5\text{V}$.

   a) The amount of charge q stored by the capacitor.

   b) The amount of energy U stored by the capacitor.

2. Consider a parallel plate capacitor where the gap distance is changed by an amount Δh. Use a linear approximation to show that $\frac{\Delta C}{C} = -\frac{\Delta h}{h}$.

   Hint: Start with $\Delta C = \Delta h \frac{dC}{dh}$.

3. Your finger tips are mostly made of water, and the touchscreen on your phone uses tiny coplanar capacitive sensors to detect your fingers. Consider a parallel plate capacitor with air inbetween the plates. How many times greater will the capacitance become if the gap is filled with water instead of air?

8 Inductive Sensors

Inductive transducers use **electromagnetic inductance** to measure distance and to detect ferromagnetic materials. That is, the position of a material containing iron or nickel can be easily detected using a coil of wire known as an **inductor**.

Notes: Inductors are typically used in AC circuits, which is the subject of Chapters 11 and 12.

8.1 Electromagnetic Inductance

According to Maxwell's equations, a changing magnetic field induces an electric field. Magnetic fields can also be induced my driving current through a wire. Changing the current changes the magnetic field, which induces an electric field. The induced electric field creates a voltage that opposes the change in current. This complicated phenomenon is neatly summarized by Lenz's Law, which says

Notes: Inductance has unit of Henry: $1 \text{ H} = 1 \text{ kg m}^2 \text{ s}^{-2} \text{ A}^{-2}$.

$$V(t) = -L\frac{dI}{dt}, \tag{8.1}$$

where V is the induced voltage, L is the inductance of the wire or "inductor", and I is the current.

The inductance L depends on the geometry of the inductor. Similar to the parallel plate capacitor, there is an idealized model for an inductor known as a **solenoid**. Illustrated in Fig. 8.1, the solenoid is a cylinder with a wire wrapped around it. The inductance of a solenoid is

$$L = \frac{\mu N^2 A}{l}, \tag{8.2}$$

where μ is the magnetic permeability, N is the number of complete turn the wire makes as it wraps around the cylinder, A is the cross-sectional area of the cylinder, and l is the length.

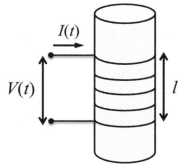

Figure 8.1: A solenoid with $N = 5$ windings.

The magnetic permeability μ depends on what material the cylinder is made of:

- For vacuum, air, or most non-ferromagnetic materials $\mu_0 = 12.6 \times 10^{-7}$ N/A^2.

- For iron $\mu_{Fe} = 6.3 \times 10^{-3}$ N/A^2.

- For carbon steel $\mu_{steel} = 1.26 \times 10^{-4}$ N/A^2.

Thus, the inductance can be changed dramatically by introducing a ferromagnetic material such as steel. This is physical principal behind most inductive transducers.

8.2   Linear Variable Displacement Transducer (LVDT)

The **LVDT** is used to measure the linear displacement of a mechanical part. In its most basic form, the LVDT is a hollow solenoid with a ferromagnetic rod, as illustrated in Fig. 8.2. Inserting or removing the rod by a distance Δx changes the inductance by an amount ΔL. An external circuit can then be used to convert the inductance to an analog voltage. The result is an analog transducer that converts displacement Δx to a voltage V.

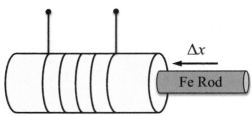

Figure 8.2: An LVDT typically consists of a solenoid with a rod made of some ferromagnetic material.

8.3   Inductive Tachometer

Notes: We will revisit the tachometer and counter-timer in Lecture 17 on digital electronics.

A **tachometer** is a sensor that measures the angular speed ω, or revolutions per minute (RPM), of a rotating mechanical part. The inductive tachometer uses an inductive coil to sense a nearby ferromagnetic part, such as the teeth on a gear. As the part rotates, the gear teeth or a small magnet mounted to the part induces a voltage in the coil of wire. The result is a series of voltage pulses that are detected and processed by a **counter-timer circuit**, which calculates the angular speed (typically in RPMs).

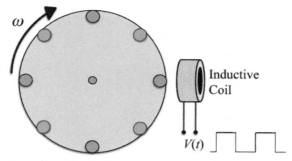

Figure 8.3: An illustration of an inductive tachometer used to measure angular speed ω of a rotating mechanical part.

Exercises 8:

1. A coil in an electric motor has an inductance of 0.5 mH (milli-Henry). The current in the coil is decreasing at a rate of 5.2×10^5 A/s. What is the induced voltage?

2. The current in an inductor oscillates sinusoidally, such that $I(t) = |I|sin(\omega t)$. Derive a formula for the induced voltage $V(t)$.

3. A solenoid consists of a wire wrapped around a 14mm long hollow tube with $N = 100$ windings, which gives it an inductance of $10^{-4}H$. What will its inductance be if the hollow center is replaced with an iron core?

9 Sensor Calibration

As discussed in previous lectures, an analog sensor converts a physical stimulus into a voltage that can easily be measured by a digital multimeter (DMM) or an analog-to-digital converter. A **calibration equation** is then used to determine the physical value from the voltage.

Sensor calibration is an experimental procedure that determines the mathematical relationship between sensor output and the physical stimulus being measured. Specifically, the sensor output is compared to a known standard. In this lecture, we will outline three different types of sensor calibration.

9.1 Two-point Linear Calibration

Most analog sensors have a linear output, which means the output voltage of the analog transducer is linearly related to the physical stimulus. For example, an analog temperature sensor might have a calibration formula

$$T = aV_{out} + b \tag{9.1}$$

where a and b are the calibration constants. Specifically, the slope a is is the inverse of the sensitivity $k = \frac{dV_{out}}{dT} = 1/a$, and the intercept b is often referred to as the "offset".

Importantly, a line is uniquely determined by two data points, as shown in Fig. 9.1. Thus, the calibration constants a and b can be determined by measuring the output voltage at two known temperatures. It is common to use boiling water and ice water for the known standard temperatures. The SI units of temperature have been judiciously defined such that ice water is at 0°C or 273 K, and boiling water is at 100°C or 373 K.

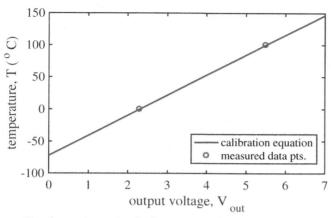

Figure 9.1: Two data points uniquely determine the linear calibration equation.

Example 9.1:

A grad student inherits an unknown RTD mounted as R_S in the circuit shown below. Unfortunately, all the documentation on the system has been lost. Assuming the output is linear, the grad student decides to perform a two-point calibration. The student places the RTD in an ice bath and records $V_{out} = 2.3$V, then places it in a beaker full of boiling water and records $V_{out} = 5.5$V, as plotted in Fig. 9.1

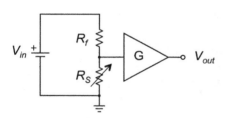

Determine the calibration constants a and b. (Use units of volts and degrees Celsius.)

Solution: We know the temperature of an ice bath is $T_C = 0°$C, and boiling water is $T_H = 100°$C. Thus, we can use these known temperatures and measured voltages to write the following equations.

$T_C = aV_C + b$

$T_H = aV_H + b$

Solving the two equations for the two unknowns, a and b, yields the following.

$$a = \frac{T_H - T_C}{V_H - V_C}$$

$$b = \frac{(T_C - T_H)V_C}{V_H - V_C}$$

Substituting the given values for V_C and V_H yields $a = 31.25°C/V$ and $b = -71.9°C$.

9.2   Multiple Point Calibration

In a **multiple point calibration**, the sensor output is measured for more than two known values. For example, an analog pressure sensor can be connected to a manometer, as shown in Fig. 9.2(a). Using the resultant data shown in Fig. 9.2(b), the pressure measured by the manometer is plotted as a function of the sensor output voltage, and computer software is used to determine a line of best fit, where a and b will be the "fitting parameters". (We will discuss this in greater detail in the next lecture on curve fitting.)

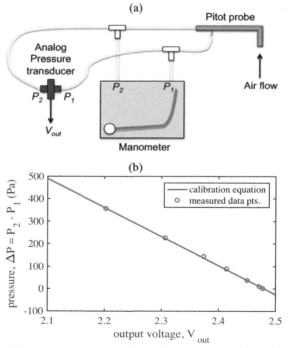

Figure 9.2: (a) A manometer can be used as a known standard for calibrating an analog pressure sensor. (b) A linear curve fit is applied to the multi-point calibration data..

9.3 Non-linear Calibration

In many cases, the output voltage is related to the physical stimulus by a **non-linear calibration formula**. For example, consider the thermistor in a voltage divider circuit shown in Fig. 9.3. The output voltage changes with the resistance of the thermistor via the voltage divider equation, Eq. (4.1), and the resistance of the thermistor is related to its temperature via the Steinhart equation, Eq. (6.2). These are both highly non-linear formulas. However it is still possible to do a simple 2-point calibration, as we will see in the following example. It is also possible to perform a multi-point calibration and apply a non-linear curve fit to the data.

Example 9.2:
A thermistor is connected in a voltage divider circuit, as shown in Fig. 9.3. An output voltage of $V_C = 0.74V$ is measured in an ice bath, and a voltage of $V_H = 4.15V$ is measured in boiling water.

a) Determine the resistance R_C of the thermistor in the cold ice bath and the resistance R_H in the hot boiling water.

Solution: First, the voltage divider equation is used to determine the resistances of the thermistor from the given voltages.

$$V_C = \frac{R_2}{R_C + R_2} V_{in}$$

$$V_H = \frac{R_2}{R_H + R_2} V_{in}.$$

Next, we solve for R_C and R_H.

$$R_C = R_2(V_{in}/V_C - 1).$$
$$R_H = R_2(V_{in}/V_H - 1)$$

Substituting in the givens yields $R_C = 27.06$ kΩ and $R_H = 0.963$ kΩ.

b) Use the known temperature of the ice bath and boiling water and the respective resistances from part a) to determine the constants in the Steinhart formula, A and B. Assume that $C \approx 0$, and $R_{ref} = 10k\Omega$.

Solution: The Steinhart equation relates the resistances R_C and R_H

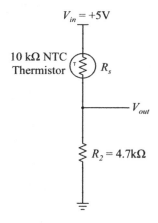

Figure 9.3: A thermistor is connected in a voltage divider to create an analog temperature sensor.

to the corresponding temperatures $T_C = 273K$ and $T_H = 373K$. Assuming that the calibration constant $C = 0$, we can write a system of two equations and two unknowns.

$$T_C = \left[A + B \ln \left(\frac{R_C}{R_{ref}}\right)\right]^{-1}$$

$$T_H = \left[A + B \ln \left(\frac{R_H}{R_{ref}}\right)\right]^{-1}.$$

Solving for A and B and using our answer from part a yields $A = 0.00347K^{-1}$ and $B = 0.000294K^{-1}$.

Exercises 9:

1. An analog temperature sensor has a linear output with calibration constants $a = 20°C/V$ and $b = -6.1°C$. Calculate the temperature if $V_{out} = 3.2V$.

2. A MEMS accelerometer outputs a voltage that is linearly related to acceleration, such that $a = AV_{out} + B$, where a is acceleration and A and B are calibration constants. A 2-point calibration is performed by rotating the axis of the accelerometer relative to gravity. When it is oriented parallel to gravity, the acceleration $a = 1g$ and the output voltage $V_{down} = 2.61V$. When it is oriented in the opposite direction, the acceleration $a = -1g$ and the output voltage $V_{up} = 1.42V$. Determine the calibration constants A and B. (Be sure to include units.)

3. An RTD is connected in a Wheatstone bridge circuit, as shown in the adjacent circuit drawing. In this circuit, $V_{in} = 5V$, $R_1 = R_2 = R_3 = 100\Omega$, and R_4 is the RTD. A 2-point calibration is performed, yielding $V_{out} = 0.009V$ in ice water at $T = 0°C$, and $V_{out} = -0.581V$ in boiling water at $T = 100°C$.

   (a) Use the Wheatstone bridge equation to calculate the resistance of the RTD in boiling water and in ice water.

   (b) Use the RTD equation to calculate the temperature coefficient of resistance (TCR), α_T in units of $1/°C$.

   (c) Do a quick Internet search. Based on the TCR you calculated in Part b, is the RTD made of platinum, tungsten, nickel, or copper?

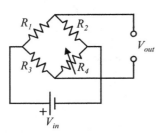

Figure 9.4: An RTD is connected in a Wheatstone bridge circuit.

10 Data Processing (Part I)

Now that we have a firm handle on sensors and electronics, we are ready to begin collecting data and performing measurements. But remember, collecting data and performing measurements only amounts to a single step in the scientific method. Equally important is how you process and interpret the data.

In this lecture, we will look at some basic techniques for processing data and comparing it to hypotheses and theoretical predictions.

10.1 The Mean and Standard Deviation

The scientific method dictates that an experimental result must be repeatable. That is, a scientist or engineer must repeat an experiment many times and consistently observe the same result for it to be credible. This is typically quantified using the mean and standard deviation.

The **mean** or **average** $\bar{x}$ is calculated by summing all the measured values x_i and dividing by the number of data points N,

$$\bar{x} = \sum_{i=1}^{N} \frac{x_i}{N}. \tag{10.1}$$

The **standard deviation** s quantifies how repeatable the result is. It is calculated by the summation

$$s = \sqrt{\sum_{i=1}^{N} \frac{(x_i - \bar{x})^2}{N}}. \tag{10.2}$$

Notes: In Matlab, the mean can be calculated using the mean() function, and the standard deviation can be calculated using the std() function.

A large standard deviation indicates a large spread in the measured values.

10.2   The Method of Least Squares

In most experiments, a dependent variable y is measured as a function of an independent variable x. The dependent variable y is then plotted as a function of the independent variable x. Hopefully, this reveals or confirms some mathematical relationship between the two variables.

The simplest mathematical relationship between two variables is the linear equation

$$y = mx + b \tag{10.3}$$

where m is the slope, and b is the intercept. As we discussed in the previous lecture, two data points can be used to uniquely determine the slope and intercept of the line. For more than two data points, the **method of least squares** can be used to determine values for m and b that best fit the data.

The method of least squares works by minimizing an error function, which is defined as the sum of the squares of the residuals

$$S = \sum_{i=1}^{N} [y_i - (mx_i + b)]^2, \tag{10.4}$$

hence the name "least squares". Illustrated in Fig. 10.1, a residual is the difference between a measured value y_i and the value predicted by the linear equation $y = mx_i + b$.

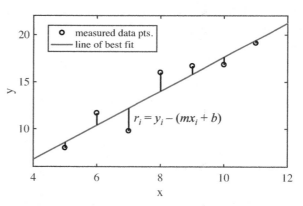

Figure 10.1: Data points are plotted with a line of best fit. The vertical distances between data points and the line are known as the "residuals".

We wish to find values for m and b that minimize the error S. To do this, we simply take the partial derivatives of S with respect to m and b and set them equal to zero. This yields

$$\frac{\partial S}{\partial m} = -2 \sum_{i=1}^{N} x_i(y_i - mx_i - b) = 0, \tag{10.5}$$

$$\frac{\partial S}{\partial b} = -2 \sum_{i=1}^{N} (y_i - mx_i - b) = 0. \qquad (10.6)$$

Solving for m and b gives us

$$m = \frac{\sum y_i (x_i - \bar{x})}{\sum (x_i - \bar{x})^2}, \qquad (10.7)$$

$$b = \bar{y} - m\bar{x}. \qquad (10.8)$$

where $\bar{x}$ and $\bar{y}$ are the average of the measured values of x and y, respectively.

The method of least squares does not only apply to linear equations. It can also be applied to fit a wide variety of analytic functions. There are many software packages that can apply both linear and non-linear curve fits. For example, Matlab has the "polyfit()" function, which can be used to fit data with an nth degree polynomial. There is also a "fit()" command available for Matlab that contains an extensive library of analytic functions, which can be used to fit measured data.

Notes: Equation (10.8) can be rewritten as $\bar{y} = m\bar{x} + b$, which means that the line of best fit always passes through the average value of x and y, or the "center of mass" of the data points.

10.3   Parameter Extrapolation

Curve fitting is commonly used for performing multi-point calibrations, as discussed in the previous lecture. Additionally, it can be used to compare a measured data set to a theoretical formula. For example, consider Galileo's inclined plane experiment, which was presented in Section 1.3. The theoretical position vs. time trajectory for the ball rolling down the inclined plane is given by

$$x(t) = \frac{1}{2} \left(\frac{5}{7} g \sin \theta \right) t^2 + v_0 t + x_0. \qquad (10.9)$$

To test or "experimentally validate" this theory, we would apply a quadratic curve fit

$$x_{fit}(t) = p_1 t^2 + p_2 t + p_3, \qquad (10.10)$$

where p_1, p_2, and p_3 are the fitting parameters, which can be related to the experimental parameters. Comparing Eqs. (10.9) and (10.10), we see that

$$p_1 = \frac{5}{14} g \sin \theta. \qquad (10.11)$$

As you will see in the following example, the fitting parameters p_1, p_2, and p_3 can be used to extrapolate values for g, v_0, and x_0.

Example 10.1:

Consider the data from Example 1.1, where a cue ball was released from a state of rest and rolled down an inclined plane with an angle $\theta = 4.6°$, and the time t it takes the ball to roll a given distance x was measured.

distance, x (m)	time, t (s)
0	0
0.1	0.646
0.2	0.915
0.3	1.120
0.4	1.294
0.5	1.446

Use the "fit()" function in Matlab to apply a quadratic curve fit to the data shown above. Then, use the calculated fitting parameters to extrapolate values for g, v_0, and x_0.

Solution: Using Eq. (10.11), we see that $g = \frac{14p_1}{5\sin\theta}$. The fitting parameter p_1 is obtained from the fit() function in lines 8 and 9 of the script below.

Matlab Script:

```
1 -    clc
2 -    close all
3
4 -    x=[0 10 20  30  40  50]*0.01;  %meters
5 -    t=[0    0.646   0.915   1.120   1.294   1.446]; %seconds
6 -    theta=4.6; %degrees
7
8 -    f=fit(t',x','poly2')
9 -    p=coeffvalues(f);
10
11 -   x0=p(3) %extrapolated initial position, m
12 -   v0=p(2) %extrapolated initial velocity, m/s
13 -   g=14*p(1)/(5*sind(theta))   %extrapolated g, m/s^2
```

Matlab Output:

```
        Linear model Poly2:
        f(x) = p1*x^2 + p2*x + p3
        Coefficients (with 95% confidence bounds):
          p1 =        0.239  (0.2374, 0.2405)
          p2 =     6.858e-05  (-0.002258, 0.002395)
          p3 =     3.714e-05  (-0.00076, 0.0008343)

    p =

      2.3898e-01   6.8583e-05   3.7137e-05
```

Matlab Output (continued):

```
    x0 =

        3.7137e-05


    v0 =

        6.8583e-05


    g =

        8.3434e+00
```

Thus, the extrapolated value for the acceleration of gravity is $g = 8.34$ m/s^2, which is slightly lower than expected due to friction. The extrapolated values of x_0 and v_0 are approximately zero, as we expect.

10.4   *Power Laws and Log-log Plots*

Many phenomena in nature exhibit **power law scaling**, where parameters scale with one another via the formula

$$y = kx^n, \tag{10.12}$$

where k is some constant coefficient and n is the exponent or "power" in the power law (i.e., "x to the nth power").

Unless $n = 1$, a power law is a non-linear function. However, any power law can be linearized by plotting it on a log-log scale. That is, we can transform data by defining new variables $x' = \log(x)$ and $y' = \log(y)$. Taking the logarithm of both sides of Eq. (10.12) and applying log identities, we see that

$$\log(y) = n\log(x) + \log(k), \tag{10.13}$$

which can be written in terms of the new variables as

$$y' = nx' + \log(k). \tag{10.14}$$

Thus, any measured data which exhibits power law scaling will appear linear on a log-log scale with a slope of n and an intercept of $\log(k)$.

A famous example of a power law is Kepler's third law for orbital mechanics

$$T = \sqrt{\frac{4\pi^2}{GM}}R^{3/2}, \tag{10.15}$$

Notes: This type of analysis is very common in fluid mechanics and heat transfer. For example, see Section 17.1 of S. Paolucci's *Undergraduate Lectures on Heat Transfer.*

where T is the orbital period and R is the semimajor axis of the elliptical orbit. Put into words, planets that are further away from the sun take longer to complete one full orbit. Measured values of $log_{10}(T)$ and $log_{10}(R)$ are plotted in Fig. 10.2 with a linear curve fit. The slope of the line is exactly 3/2, and the intercept is related to the mass of the sun.

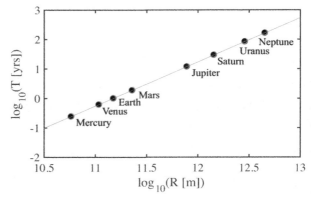

Figure 10.2: A log-log plot of the planet's orbital period vs. semimajor axis linearizes the data with a slope of 3/2.

10.5   The Danger of Arbitrary Curve Fits

Most modern software packages offer a wide variety of functions that can be used to curve fit data. However, it is important not to arbitrarily choose a fitting function. Rather, a fitting function should be chosen based upon some scientific theory. In Example 10.1, we chose to fit the inclined plane data with a quadratic curve, because Newtonian mechanics—namely Eq. (10.9)—tells us that the data should be quadratic. We could have arbitrarily chosen an exponential function or a sine wave, but those equations are not consistent with the established theoretical trajectory. This can result in drastically incorrect predictions. For example, the sine wave predicts that the ball would turn around and roll *back up the ramp* after a certain amount of time.

Notes: Linear curve fits are usually a safe choice, because most analytic functions can be approximated by a line over a small enough domain, according to Taylor's theorem.

Polynomial curves are the most problematic arbitrary curve fits, and they are commonly used by uninformed undergraduate students (and also uninformed professionals). Consider the two-point calibration in the previous lecture, where two data points were used to determine the equation of a line that passed perfectly through both data points. The line was a polynomial of degree $n = 1$, but this can be generalized to any n^{th} degree polynomial, where a polynomial of degree n can be uniquely defined by $n + 1$ data points.

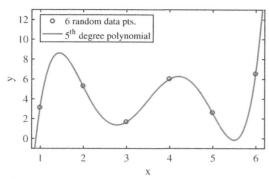

Figure 10.3: A 5th order polynomial passes perfectly through 6 random data points.

Figure 10.3 shows six data points that were created using a random number generator. A polynomial of degree $n = 5$ was used to curve fit the data. Note that the curve passes *perfectly* through all six data points. Does this mean that the polynomial can be used to predict the location of a 7th randomly generatred data point? Absolutely not. The data is totally random. This illustrates the detriment of fitting data with an arbitrary curve. One can easily be mislead into thinking that data follows a certain mathematical trend when it clearly does not.

Exercises 10:

1. Derive Eqs. (10.7) and (10.8) by algebraically solving Eqs. (10.5) and (10.6).

2. An analog pressure transducer is calibrated using the setup shown in Fig. 9.2(a), and the resultant calibration data is shown in the table adjacent to this problem.

    (a) Plot the calibration data with a linear curve fit, similar to Fig. 9.2(b).

    (b) Based on the linear curve fit, what are the calibration constants a and b for the calibration equation $P = aV_{out} + b$? (Be sure to include units.)

P (Pa)	V_{out} (V)
0	2.477
87.2	2.541
142	2.581
224	2.648
354	2.752

3. The orbital periods T and semimajor axes R for Jupiter's four largest moons are shown in the table below. Use the data to make the following.

moon	period, T (days)	semimajor axis, R (km)
Io	1.77	421700
Europa	3.55	671034
Ganymede	7.15	1070412
Callisto	16.69	1882709

(a) Make a plot of the raw data, T as a function of R, using Matlab.

(b) Make a plot of $\log_{10}(T)$ vs $\log_{10}(R)$, similar to Fig. 10.2.

(c) Apply a linear curve fit to the log-log transformed data. What is the slope? Is it consistent with Kepler's third law?

11 Alternating Current (AC)

So far, we have only considered **direct current (DC)** electronics, where the current and voltage are both constant in time. We will now consider **alternating current (AC)**, where the applied voltage oscillates in time.

11.1 AC Sine Waves

Electrical energy is primarily produced at power plants by generators and distributed through a network of high voltage transmission lines. Generators use rotating electromagnetic inductors to produce voltage and current that oscillates in time. In North America, the resultant electricity that comes from our wall outlets oscillates as a sine wave with a frequency of $f = 60$ Hz and a peak amplitude of $|V| = 170$ V. This particular **AC voltage waveform** is plotted in Fig. 11.1.

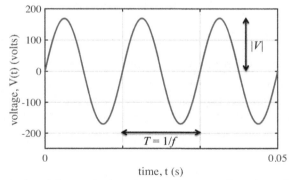

Figure 11.1: A plot of the standard 60 Hz AC voltage waveform for wall outlets in North America.

Listed below are several parameters that are used to characterize and define an AC waveform.

- **Frequency** f is the number of full cycles per second. It has units of "Hertz", where 1 Hz = 1 cycle/second.

- **Angular frequency** ω is the frequency in "radians per second".

Note that there are 2π radians per cycle, so $\omega = 2\pi f$.

- **Period** T is the time it takes to complete one full cycle of the sine wave. It can be calculated as $T = 1/f$.

- **Peak amplitude** $|V|$ is the maximum value attained by the AC sine wave minus the average value of the wave. It defines how "tall" the sine wave is.

- **Peak-to-peak amplitude** V_{pp} is the vertical distance between the maximum and minimum value of the sine wave, such that $V_{pp} = 2|V|$.

- **Root-mean-square (RMS) Amplitude** V_{RMS} is defined by the formula

$$V_{RMS} = \sqrt{\frac{1}{T} \int_0^T [V(t)]^2 \, dt} \qquad (11.1)$$

For a sine wave, $V_{RMS} = |V|/\sqrt{2}$.

11.2 Amplitude and Phase

In most cases, the AC voltage waveform $V(t)$ is given by the function

$$V(t) = |V| \sin(\omega t), \qquad (11.2)$$

where $|V|$ and ω are known parameters. Applying this AC voltage to the circuit results in an oscillating current

$$I(t) = |I| \sin(\omega t - \phi), \qquad (11.3)$$

where $|I|$ is the **amplitude** of the current and ϕ is the angular **phase shift** of the current. Note that the current and voltage both oscillate at the same frequency, but they have different amplitudes and are shifted by a phase angle ϕ. Much of this and the next lecture will be devoted to techniques for calculating and measuring the amplitude and phase of the current.

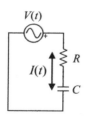

Figure 11.2: An AC voltage source is used to power an RC circuit.

Notes: Multiplying ω in radians/second times Δt in seconds gives ϕ in units of radians.

11.3 AC Circuits

Shown in Fig. 11.2, an AC voltage source drives an RC circuit with a sinusoidal voltage $V(t)$. This causes current $I(t)$ to oscillate back-and-forth through the circuit, periodically charging and discharging the capacitor. As previously stated, the current will have an amplitude $|I|$ and phase ϕ. The phase angle ϕ corresponds to a time lag Δt between the two waveforms, where $\phi = \omega \Delta t$.

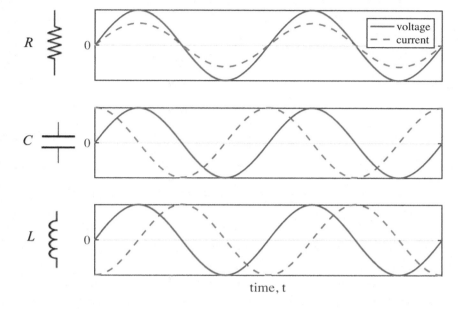

Figure 11.3: Resistors, capacitors, and inductors result in different phase shifts.

An AC circuit can be constructed using any combination of resistors, capacitors, and inductors. The amplitude of the current and its phase depend on the values of the chosen circuit elements. We will discuss this in detail in the next lecture. For now, you should recognize the following:

- In a **resistor**, the phase shift $\phi = 0$.

- In a **capacitor**, the current leads the voltage with a phase shift $\phi = -90° = -\pi/2$.

- In an **inductor**, the current lags behind the voltage with a phase shift $\phi = +90° = \pi/2$.

This is all illustrated in Fig. 11.3. Trace the current waveforms with a pen. For a capacitor, you will see that the current crosses zero and peaks before the voltage, meaning that it *leads* the voltage.

Notes: "ELI the ICE man" is popular mnemonic used to remember the phase shifts of inductors and capacitors.

- **ELI** - For an inductor L, the voltage E leads the current I.

- **ICE** - For a capacitor C, the current I leads the voltage E.

Example 11.1:

An electric vehicle (EV) uses an AC induction motor. The motor's stator can be modeled by the RL circuit shown below.

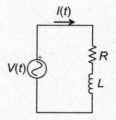

The stator circuit is driven by a sinusoidal AC voltage with a frequency $f = 50\text{Hz}$ and a peak amplitude $|V| = 300\text{V}$. The resultant current has a peak amplitude of $|I| = 75\text{Amps}$ with a phase $\phi = 86°$ lagging behind the voltage.

Use the "yyaxis left" and "yyaxis right" commands in Matlab to make a plot of both the current and voltage as a function of time on the same graph. Voltage should go on the left y-axis, and current should go on the right y-axis. (Be sure to include units.)

Matlab Script:

```
clc
close all

%Givens
f=50;          %frequency, Hz
T=1/f;         %period, s
w=2*pi*f;      %angular frequency, rad/s
phi=86*pi/180; %phase shift, rad
Vamp=300;      %voltage peak amplitude, Volts
Iamp=75;       %current peak amplitude, Amps

t=linspace(0,3*T,1000); %plot 3 full periods of oscillation

V=Vamp*sin(w*t); %voltage vector, same length as t
I=Iamp*sin(w*t-phi); %current vector, same length as t

figure(1) % new figure

yyaxis left
plot(t,V,'linewidth',2)
xlabel('time, t [s]')
ylabel('voltage, V(t) [V]','fontSize',16) % left y-axis

yyaxis right
plot(t,I,'linewidth',2)
ylabel('current, I(t) [A]','fontSize',16) % right y-axis
set(gca,'fontSize',16)
```

Matlab Output:

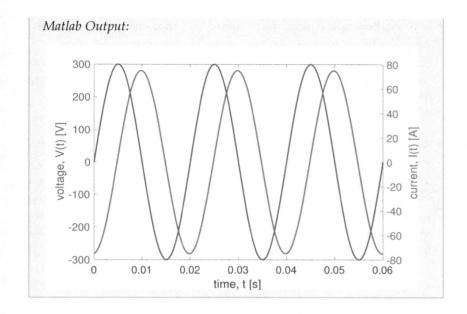

11.4   Lab Equipment and AC Measurements

There are several important pieces of lab equipment that are commonly used for developing and testing AC circuits.

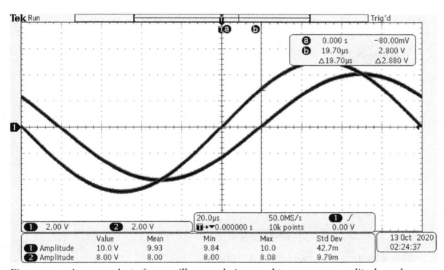

Figure 11.4: A screen shot of an oscilloscope being used to measure amplitude and phase.

- A **function generator** is used to produce an AC voltage waveform. It can produce sine waves and other periodic functions, such as square waves and ramp waves, with almost any desired amplitude and frequency.

- An **oscilloscope** is used to examine and measure AC waveforms. Specifically, it can be used to measure amplitude and phase. A screenshot from a digital oscilloscope is shown Fig. 10.4. The two vertical lines are known as "cursors", and they are used to measure the time lag between the two waveforms Δt. The phase angle $\phi = \omega \Delta t$.

- **Coaxial cable** and **BNC connectors** are typically used to make electrical connections in AC circuit measurements. Shown in Fig. 10.5, a central wire, or "inner conductor" carries the signal $V(t)$. A woven cable forms a coaxial cylindrical "jacket" around the signal in the center. The jacket and inner conductor are separated by white polymer insulation. The outer jacket is always grounded and shields the signal from **electromagnetic interference (EMI)**.

Notes: Use a pen to trace out the current loop formed by the coaxial cable in Fig. 11.5.

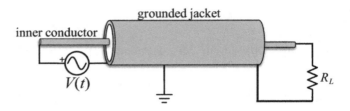

Figure 11.5: An illustration of how a coaxial cable is used in an AC circuit.

Exercises 11:

1. Combine Eqs. (11.1) and (11.2) and integrate to show that $V_{RMS} = |V|/\sqrt{2}$ for a sine wave.
   (Hint: Use the half angle formula $2\sin^2(A) = 1 - \cos(2A)$.)

2. An AC voltage source is used to drive the RC circuit shown in Fig. 11.2 with a frequency $f = 20,000$Hz and a peak amplitude $|V| = 10$V. Calculate the following:

   (a) The period T.
   (b) The angular frequency ω.
   (c) The peak-to-peak amplitude V_{pp}.
   (d) The RMS amplitude V_{RMS}.
   (e) The current and voltage are separated by $\Delta t = 7\mu s$. Calculate the phase angle ϕ.
   (f) Do you expect the current to lead or lag behind the voltage?
   (g) Sketch the current and voltage waveforms on the same plot together. Use a dotted line or colored pens to differentiate the two.

3. An AC voltage source is used to drive the RC voltage source shown in Fig. 11.2 with a frequency $f = 20,000$Hz and a peak amplitude $|V| = 10$V. The resultant current $I(t)$ leads the current by a phase of $\phi = -42°$ and has an amplitude $|I| = 23$mA.

Use the "yyaxis left" and "yyaxis right" commands in Matlab to make a plot of both the current and voltage as a function of time on the same graph. Voltage should go on the left y-axis, and current should go on the right y-axis. (Be sure to include units.)

12 AC Circuit Analysis

In the previous lecture, we introduced AC sine waves and the concept of amplitude and phase. This lecture will present a mathematical approach for theoretically predicting the amplitude and phase in an AC circuit. The math presented in this chapter is formidable, but it is applicable to a wide range of problems involving waves, vibrations, and oscillations, so it is worth the time to learn it.

12.1 Complex Numbers

Complex numbers are numbers that have a real part and an imaginary part. They can be written as

$$z = a + ib, \tag{12.1}$$

where $i = \sqrt{-1}$, a is the real part of z, b is the imaginary part of z. The real and imaginary parts can be plotted as Cartesian coordinates, as shown in Fig. 12.1. This figure is known as the **complex plane**.

It is common to use the notation $Re(z) = a$ and $Im(z) = b$, but a complex number z can also be expressed in polar form

$$z = |z|(\cos \phi + i \sin \phi), \tag{12.2}$$

where

$$|z| = \sqrt{Re(z)^2 + Im(z)^2}, \tag{12.3}$$

and

$$\phi = \arctan \left[\frac{Im(z)}{Re(z)} \right]. \tag{12.4}$$

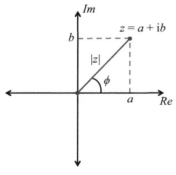

Figure 12.1: A complex number $z = a + ib$ is plotted in the complex plane.

Because $e^{i\phi} = \cos \phi + i \sin \phi$, Equation (12.2) can be expressed in the remarkably simple form

$$z = |z|e^{i\phi}, \tag{12.5}$$

where e is Euler's number. Thus, we see the utility of using complex numbers. A single complex number has two degrees of freedom that allow us to keep track of both amplitude and phase. Specifically, the magnitude $|z|$ is used to represent the amplitude of a wave, while the angle ϕ is used to represent phase.

Furthermore, Eq. (12.5) can be used to express the AC voltage waveform equation as $V(t) = Im\left[|V|e^{i\omega t}\right]$. In the subsequent sections, we will omit the "$Im[]$" and simply write the AC voltage sine wave as

$$V(t) = |V|e^{i\omega t}, \qquad (12.6)$$

and the AC current sine wave as

$$I(t) = |I|e^{i\omega t - \phi}. \qquad (12.7)$$

This form is particularly amenable algebraic manipulation, as multiplication and division reduces to adding or subtracting exponents. We may then take the imaginary part when we have finished our algebra to retrieve the original sinusoidal function and eliminate the imaginary number i.

12.2 Complex Impedances

As discussed in the previous lecture, resistors, capacitors, and inductors do not only affect the magnitude of the current $|I|$, but they also affect the phase ϕ of the current relative to the voltage. Thus, we define a complex impedance Z that is used in place of resistance in Ohm's law, such that $V(t) = I(t)Z$, where Z contains information about *both* **amplitude** and **phase**.

Notes: The complex impedances all have units of Ohms

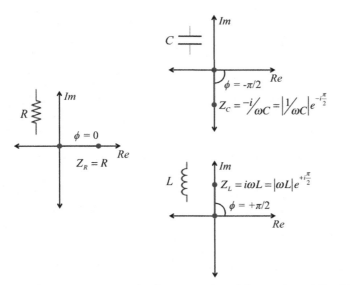

Complex Impedances

- Resistor: $Z_R = R$
- Capacitor: $Z_C = \frac{-i}{\omega C}$
- Resistor: $Z_L = i\omega L$

Figure 12.2: Resistors, capacitors, and inductors result in different phase shifts. This information is encoded in their complex impedances.

The complex impedances for a resistor, capacitor, and inductor are shown in Fig. 12.2. Note that the polar angle in the complex plane corresponds to the phase shifts listed in the previous lecture for each circuit element. Consider, for example, the current in the capacitor

$$I(t) = \frac{V(t)}{Z_c} = \frac{|V|e^{i\omega t}}{\frac{1}{\omega C}e^{-i\pi/2}}. \tag{12.8}$$

Simplifying, the current becomes

$$I(t) = |V|\omega C e^{i(\omega t + \pi/2)}, \tag{12.9}$$

where the amplitude of the current is $|I| = |V|\omega C$ and the phase angle $\phi = \pi/2$.

These impedances can also be combined in series and in parallel, similar to resistors, to create an equivalent impedance Z_{eq}. The equivalent impedance, like any complex number, can be expressed in the form

$$Z_{eq} = |Z_{eq}|e^{i\phi}. \tag{12.10}$$

Applying Ohm's Law to calculate current yields

$$I(t) = \frac{|V|}{|Z_{eq}|}e^{i\omega t - \phi}. \tag{12.11}$$

Here we see that the amplitude of the current $|I| = |V|/|Z_{eq}|$ and the phase angle is simply the polar angle of Z_{eq} in the complex plane.

Notes: Inductors and capacitors can be treated as a simple resistors with an appropriate impedance, and the AC circuit analysis becomes a straightforward application of Kirchhoff's circuit laws.

12.3   AC Circuit Analysis

The complex impedances listed in Fig. 12.2 make AC circuit analysis simple and mathematically elegant. Every circuit element can be treated as a simple resistor with an appropriate impedance, and the problem becomes a straightforward application of Kirchhoff's circuit laws. This process is demonstrated in the following example.

Example 12.1:

The RC circuit shown below is driven by an AC sine wave with amplitude $|V| = 1V$ and a frequency of $f = 1$ kHz. Calculate the amplitude and phase of the current if $R = 1$ kΩ and $C = 1\mu$F.

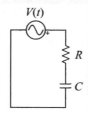

Solution: We redraw the circuit using the impedances from Fig. 12.2 and apply Kirchhoff's circuit laws.

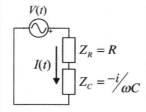

Adding the two impedances in series gives us the equivalent impedance $Z_{eq} = Z_R + Z_C = R - \frac{i}{\omega C}$.

The amplitude of the current is then $|I| = \frac{|V|}{|Z_{eq}|}$. Applying Eq. (12.3) to $|Z_{eq}|$, we have

$$|I| = \frac{|V|}{\sqrt{R^2 + (1/\omega C)^2}}.$$

Substituting in the givens, gives us $|I| = 0.988$mA.

According to Eqs. (12.10) and (12.11), the phase shift is simply the polar angle of Z_{eq}. Applying Eq. (12.4) gives us

$$\phi = \arctan\left[\frac{-1}{\omega RC}\right].$$

Substituting in the givens, gives us $\phi = -0.158$ radians or $-9°$.

Notes: Don't forget to convert Hz to radians/sec. $\omega = 2\pi f$

12.4   AC Power

Most of our modern industrial civilization is supported by tremendous amounts of electrical energy that are transported via AC transmission lines. Also, AC induction motors are used in most electric vehicles. For these reasons, it is important for engineers to understand the basics of AC electrical power. Here are several things worth noting:

- **Resistors** dissipate power via Joule heating. However, the AC power dissipated is not constant, but varies in time.

- **Capacitors** do not dissipate power via Joule heating. Rather, they store and release electrical energy, as the capacitor is subsequently charged and discharged. During this process, the power will oscillated between positive and negative values.

- **Inductors** do not dissipate power via Joule heating. Rather, they store and release electrical energy in the form of induced electromagnetic fields. This also results in positive and negative powers.

The **average power** is the metric used to quantify AC power. It is highly dependent on the amplitude and phase of the current. The average power in an inductor or a capacitor is zero.

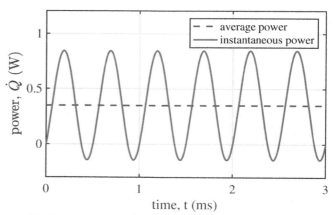

Figure 12.3: The instantaneous and average power in an AC circuit.

The **instantaneous power** in an AC circuit is simply $\dot{Q}(t) = I(t)V(t)$, which can be written using Eqs. (11.2) and (11.3) as

$$\dot{Q}(t) = |V||I| \sin(\omega t) \sin(\omega t - \phi). \qquad (12.12)$$

The instantaneous power for an RC circuit, similar to Example 12.1, is plotted in Fig. 12.3. Note that it is negative at times due to the energy stored when the capacitor is charged.

The **average power** dissipated in the circuit can be calculated using the integral formula

$$\langle \dot{Q}(t) \rangle = \frac{1}{T} \int_0^T |V||I| \sin(\omega t) \sin(\omega t - \phi)\, dt. \qquad (12.13)$$

Evaluating the integral yields the simple formula

$$\langle \dot{Q}(t) \rangle = \frac{|V||I|}{2} \cos \phi. \qquad (12.14)$$

Importantly, the average power only depends on the amplitude and phase of the voltage and current. As we saw in Example 12.1, this information is embedded in the equivalent impedance of the circuit.

Exercises 12:

1. An electric vehicle uses an AC induction motor, and the motor's stator is modeled by the same RL circuit as in Example 11.1 with $R = 0.1\Omega$ and $L = 2\text{mH}$. The RL circuit is driven by an AC voltage with a peak amplitude $|V| = 280\text{V}$ and frequency $f = 50$ Hz.

   (a) Calculate the amplitude of the current $|I|$.

   (b) Calculate the phase of the current ϕ.

   (c) Calculate the average power dissipated in the stator $\langle \dot{Q}(t) \rangle$.

2. The RC circuit shown in the adjacent drawing is driven by an AC sine wave with amplitude $|V| = 1V$ and a frequency of $f = 10\text{kHz}$.

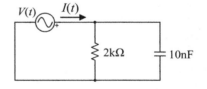

   (a) Calculate the amplitude and phase of the total current $I(t)$ drawn from the power supply.

   (b) Make a plot of the instantaneous power $\dot{Q}(t)$ vs. time t. Include a horizontal line denoting the average power $\langle \dot{Q}(t) \rangle$, similar to Fig. 12.3.

3. Consider an AC circuit containing some arbitrary network of resistors, capacitors, and inductors, which can be combined into a single circuit element with an equivalent impedance Z_{eq}, as shown in the adjacent drawing. The applied voltage has an amplitude of $|V| = 1V$, and the resultant current has an amplitude $|I| = 1\text{Amp}$.

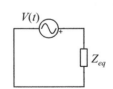

   (a) Plot the average power dissipated in the circuit as a function of the phase ϕ over a domain of $-\pi/2$ to $+\pi/2$.

   (b) What values of ϕ maximize the average power?

   (c) What values of ϕ minimize the average power?

   (d) Sketch the complex plane. Indicate the location of values of Z_{eq} that will maximize and minimize the power.

13 Analog Filters

The next few lectures will focus on acquiring and processing transient signals from analog sensors. For example, if we wish to see how fast the engine of a car heats up, we would mount an analog temperature sensor that would output some transient voltage signal $V(t)$.

13.1 Electronic Noise

In the previous two lectures, we considered voltages that oscillate as perfect sine waves. A perfect sine wave is merely a mathematical construct that does not exist in reality. Any AC sine wave measured in the real world will always have some unwanted **electronic noise** superimposed on it. An example of this is shown in Fig. 13.1. The AC sine wave, measured using an oscilloscope, appears to be a bit "fuzzy" due to electronic noise.

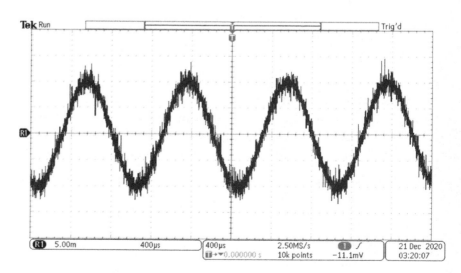

Figure 13.1: A sine wave measured on an oscilloscope exhibits noise.

Electronic noise is an unwanted oscillation in a signal, and it is typically due to two things.

- **Electromagnetic interference (EMI)** comes from surrounding electronic devices and other naturally occurring phenomena.

- **Thermal noise** comes from the random motion of electrons carrying the charge in the circuit.

13.2   Electronic Filters

Electronic filters are tertiary circuits that are used to reduce the noise in an analog measurement or signal. Filters typically block a certain range of frequencies. Shown in Fig. 13.2, we see that a **low-pass** filter passes the low frequency component of the signal and blocks the high-frequency component.

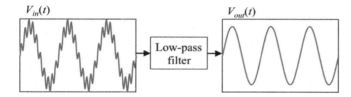

Figure 13.2: A low-pass filter passes low frequency signals and blocks high frequency noise.

There are a wide variety of filters available for conditioning signals. Analog signal conditioning is a topic in its own right, and there are entire books written on the subject. For our immediate purposes, this lecture will only serve as an overview and provide a few simple examples of common filters.

13.3   The Low-pass Filter

The **low-pass filter** passes low frequency components and blocks high frequency components. What do we mean by a "frequency component"? Think of a university marching band playing your school's fight song. Each instrument plays a certain tone or frequency. The tubas play low frequency bass, while piccolos play high frequency treble. Each instrument represents an individual frequency component, and played together, they form the melody of the song. The song you hear played by the band is merely an acoustic pressure $P(t)$, and a microphone is an analog pressure sensor that converts the pressure to a voltage $V(t)$. Feeding this signal into a low-pass filter would eliminate high frequency tones, like those produced by the piccolos.

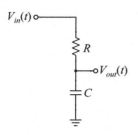

Figure 13.3: A **low-pass filter** circuit constructed with a resistor R and capacitor C.

The simplest way to construct a low-pass filter is with the RC circuit shown in Fig. 13.3. For an AC sine wave with an input amplitude $|V_{in}|$, the amplitude of the output is given by the equation

$$|V_{out}| = \frac{1}{\sqrt{1 + (\omega RC)^2}} |V_{in}|. \tag{13.1}$$

Note that the output amplitude goes to zero as $\omega \to \infty$, and $|V_{out}| \approx |V_{in}|$ for low frequencies near $\omega = 0$.

The resistance R and the capacitance C dictate the **cutoff frequency** of the circuit

$$\omega_c = \frac{1}{RC}. \tag{13.2}$$

The amplitude ratio $|V_{out}|/|V_{in}|$ is plotted in Fig. 12.4 with a vertical line at the cutoff frequency ω_c. We see that signal is significantly **attenuated** for frequencies above ω_c.

Notes: The term "cut-off" frequency is a bit misleading. The signal is not abruptly cut off at ω_c, rather it is gradually **attenuated**, as seen in Figs. 13.4 and 13.6.

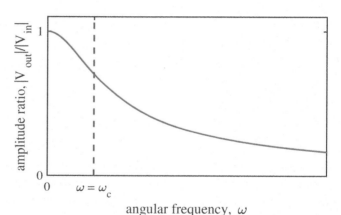

Figure 13.4: The frequency response curve of an RC low-pass filter.

13.4   The High-pass Filter

The **high-pass filter** passes high frequency components and blocks low frequency components. Using music again as an analogy, the high-pass filter would block bass and pass treble. The simplest way to construct a high-pass filter is with the RC circuit shown in Fig. 13.5. For an AC sine wave with an input amplitude $|V_{in}|$, the amplitude of the output is given by the equation

Notes: RC has units of time, and ωRC is **dimensionless** or **non-dimensional** meaning that all of the units cancel out.

$$|V_{out}| = \frac{\omega RC}{\sqrt{1 + (\omega RC)^2}} |V_{in}|. \tag{13.3}$$

As with the low-pass filter, the resistance R and the capacitance C dictate the cutoff frequency of the circuit $\omega_c = \frac{1}{RC}$. However, the signal is significantly attenuated for frequencies *below* ω_c, as shown in Fig. 13.6.

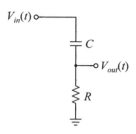

Figure 13.5: A **high-pass filter** circuit constructed with a resistor R and capacitor C.

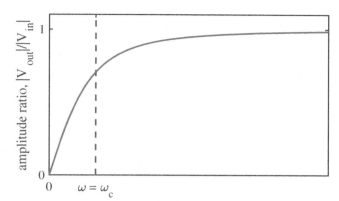

Figure 13.6: The frequency response curve of an RC high-pass filter.

13.5   Band-pass and Notch Filters

The **band-pass filter** passes signals within a certain **frequency band** that is centered at some desired frequency ω_0. The bandwidth is characterized by the **full width at half-max (FWHM)** $\Delta\omega$, which is the width of the peak at half of its maximum value. The frequency response of the band-pass filter is plotted in Fig. 13.7(a).

Notes: Band-pass filters are mostly used as radio tuners, which select the *carrier frequency* of a radio signal. Listed below are the carrier frequency bands for various forms of wireless radio communication.

- **AM Radio** 550 to 1720 kHz
- **FM Radio** 88 to 108 MHz
- **WiFi** 2.4 to 2.5 GHz
- **5G Mobile** 28 to 39 GHz

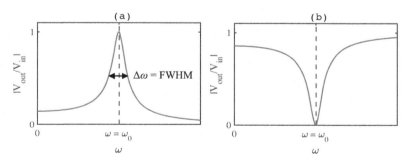

Figure 13.7: The frequency response curve of (a) a band-pass filter and (b) a notch filter.

The **notch filter** *attenuates* signals within a certain frequency band that is centered at ω_0. The notch filter also has a characteristic bandwidth $\Delta\omega$, as shown in Fig. 12.7(b). Notch filters are very common in consumer electronics, and they are used to block 60Hz electromagnetic interference produced by AC wall outlets.

Band-pass and notch filter circuits can be quite complex, and it is typically more cost effective to simply purchase one "off the shelf" from a vendor, rather than try to build one from scratch. For example, a lock-in amplifier is a high-quality band-pass filter that can be set to any frequency ω_0.

Exercises 13:

1. Show that $|V_{out}|/|V_{in}| = 1/\sqrt{2}$ at the cut-off frequency $\omega = 1/RC$ for *both* the low-pass and high-pass RC filters.

2. Consider the following problematic situations, and determine which filter would be best to remedy it (low-pass, high-pass, band-pass, notch).

    (a) A quadrocopter drone is carrying a thermistor to measure the air temperature. Electromagnetic interference (EMI) from the drone's motors results in high frequency noise > 100Hz.

    (b) You live underneath massive high-voltage transmission lines, and your stereo system constantly has a 60Hz hum.

    (c) A sensor is used to measure the blade passage rate of an aircraft propeller spinning at 1000s of RPMs. Thermal fluctuations cause the sensor voltage to slowly drift at a low frequency < 1Hz.

3. An engineer wishes to build a low-pass filter RC circuit with a cut-off frequency of 10kHz.

    (a) Determine the cut-off frequency ω_c in radians/sec.

    (b) The engineer has a small variety of resistors and capacitors on-hand in the lab. The available values of R and C are listed in the adjacent table. Sketch the circuit, and choose a value for R and for C from the table that will yield a cutoff frequency within 10% of the value you calculated in part (a).

Resistors	Capacitors
10 Ω	10 pF
100 Ω	100 pF
220 Ω	1.5 nF
1 kΩ	4.7 nF
4.7 kΩ	470 nF
10 kΩ	10 μF

14 Semiconductor Devices

So far, we have covered resistors, capacitors, and inductors, and this may have been a review of things you learned in your Physics II course. We will now move on to more complex circuit elements known as semiconductor devices. **Semiconductor devices** are made from non-metal elements, typically silicon, which only conduct electrical current under special circumstances.

This lecture will primarily focus on **diodes** and **transistors**. These circuit components can be found in nearly every consumer electronic product available in our modern world. They are also commonly used in conjunction with analog sensors, so it is important for any engineer to posess a basic understanding of their operational principles.

14.1 Diodes and LEDs

A **diode** is a device that only allows current to flow in one direction. Shown in Fig. 14.1, current flows when the diode is oriented in the **forward bias** direction, and current does *not* flow when it is oriented in the **reverse bias** direction. Thus, the diode is a **non-ohmic** device, which means that its current-voltage relation does not obey Ohm's law. This is apparent in the current voltage relation plotted in Fig. 14.2, where the current grows exponentially for positive voltages, while only a small amount of leakage current flows through the diode for negative voltages.

All diodes emit light when a positive, forward-bias voltage drives current through them. For common silicon, the emitted light is typically in the infrared spectrum, which is invisible to our human eyes. **Light emitting diodes (LEDs)** are diodes that specially designed to emit light in the visible spectrum. They are typically made from elements in groups III and V of the periodic table, such as indium (In), gallium (Ga), phosphorus (P), and arsenic (As).

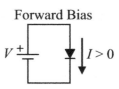

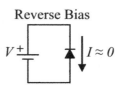

Figure 14.1: Current flows through a diode connected in forward bias (top), but not in reverse bias (bottom).

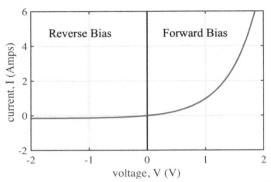

Figure 14.2: The current voltage relation for a diode clearly does not obey Ohm's law.

14.2 Transistors

A transistor is a 3-terminal device that only allows current to flow between two of the terminals when a certain voltage is applied to the third terminal. The two most common types of transistor are shown in Fig. 14.3. The **bipolar junction transistor (BJT)** and the **field effect transistor (FET)** operate based on different physical principles and have slightly different nomenclature. Within the scope of this book, there are two important things to know.

- The **BJT** only allows current I_C to flow from the collector to the emitter when a small amount of current is injected into the base.

- The **FET** only allows current I_D to flow from the drain to the source when a sufficient voltage is applied to the gate.

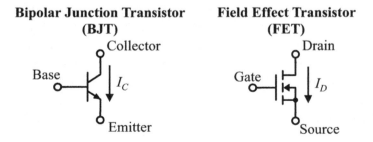

Figure 14.3: Schematic representations of a BJT (left) and an FET (right).

The operational principal is the same for both a BJT and FET: a weak, low-power signal applied to one terminal controls a large amount of current between the other two terminals. When a large amount of current is flowing through the transistor, we say that the transistor is turned ON.

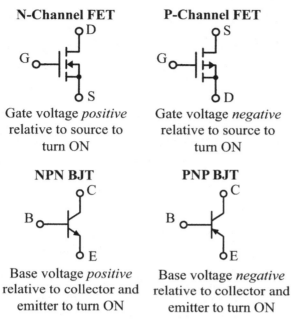

Figure 14.4: FETs can be either N-channel or P-channel (top), while BJTs can be either NPN or PNP (bottom). This determines whether the threshold voltage is positive or negative.

All transistors have a certain **threshold voltage** V_{TH} necessary to turn them on. The threshold voltage can be either negative or positive, depending on the physical make-up of the transistor. Accordingly, both FETs and BJTs can be sub-divided into two categories shown in Fig. 14.4.

14.3 Practical Uses of Transistors

Transistors have two primary functions in electronics:

- **Amplifiers** – Transistors can be used to take a weak signal and amplify it, so it has more power. (We will discuss this further in the next lecture.)

- **Logic circuits** – Transistors obey an inherent logic: The transistor is ON if a sufficient voltage is applied to the gate/collector, or the transistor is OFF if a no voltage is applied to the gate/collector. The simple logic of one transistor can be combined with other transistors to create sophisticated devices capable of complex mathematical operations. This is the basis of the modern digital computer.

In the following example, we will see how a transistor can be used as both an amplifier and a logic circuit.

Example 14.1:
An engineer wants to use a weak $0-5V$ signal from a computer
to turn a 12V electric motor on an off. The 5V output from the
computer is not powerful enough to drive the 12V motor, so an
external 12V DC power supply is used for the motor.

Solution: A field effect transistor is used to amplify the weak signal from the computer, as shown in the circuit drawing below.
This particular transistor has a threshold voltage $V_{TH} = 3V$.

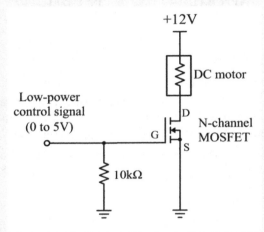

- When 0V is applied to the gate, current can not flow through
the motor, because the transistor is in the OFF state.

- When 5V is applied to the gate, the gate voltage is now positive relative to the source and above the threshold voltage
$V_{TH} = 3V$. This switches the transistor to the ON state, allowing current to flow through the motor.

Thus the transistor serves two purposes. It amplifies the weak
signal from the computer, and it allows for logical operation of
the motor.

The 10 kΩ resistor is known as a **pull-down resistor**. When the
computer is shutdown, this resistor pulls the gate voltage down
to 0V, which prevents the gate from becoming positively charged
by static electricity and randomly turning on the motor.

14.4   Integrated Circuits

The integrated circuit is one of the most important inventions in all of human history, as it is the basis for all modern digital electronics and computers. An **integrated circuit (IC)** is a small piece of ultra high purity silicon with a microscopic circuit printed on it. These IC chips can contain billions of transistors and are used to make everything from microprocessors to photo sensors in digital cameras.

The most common structure found on an IC chip is the metal-oxide-semiconductor field effect transistor (MOSFET). Shown in Fig. 14.5, an N-channel MOSFET is constructed on a piece of "p-doped" silicon that intentionally contains boron impurities. The source and drain are created by adding phosphorus or arsenic impurities to create "n-doped" regions.

Notes: As of 2018, it is estimated that humans have manufactured over 10^{22} MOSFETs worldwide.

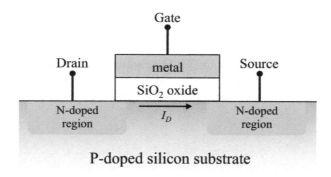

Gate

Drain

Source

metal

SiO_2 oxide

N-doped region

I_D

N-doped region

P-doped silicon substrate

Figure 14.5: A cross sectional view of the basic structure of a MOSFET.

The gate oxide layer is formed by baking the silicon in a very hot and humid furnace, causing the silicon to grow a thin layer of common glass, or SiO_2, which is an excellent dielectric. The "channel" between the source and drain becomes conductive when a voltage is applied to the gate. The gate oxide is capped with a metal contact, thus forming the "metal-oxide-semiconductor" structure at the center of the MOSFET.

Notes: MOSFETs can be fabricated with dimensions less than 10^{-7}m, many times smaller than a red blood cell.

Exercises 14:

1. Consider the adjacent circuit, where alternating current is driven through a diode and resistor.

   (a) Sketch the voltage drop across the resistor $V_{out}(t) = I(t)R$ as a function of time.

   (b) Can you think of anything that this circuit could be used for?

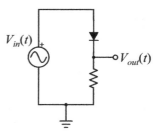

2. An engineer wishes to automate a process, where a cooling fan (driven by a 12V DC motor) is turned on when the system temperature becomes greater than a certain threshold.

(a) Sketch a simple circuit that will achieve this task. Combine the circuit from Example 14.1 with the thermistor voltage divider circuit from Example 9.2.

(b) How might you modify the circuit, so that an operator can manually adjust the threshold temperature for the fan to turn on?

3. Early digital cameras used a special integrated circuit known as a charge coupled device (CCD). Each pixel in the CCD is a small metal-oxide-semiconductor (MOS) capacitor that detects photons by converting their energy to electrical charge.

(a) In one particular prototype of a CCD, the MOS photo-detectors were square shaped with a side length of 20 μm. How many of these pixels can fit on a 2×2 cm silicon wafer?

(b) In a later version of the CCD, engineers managed to shrink the size of the MOS photo-detectors down to 5 μm. How many of these new-and-improved smaller pixels can fit on a 2×2 cm silicon wafer?

15 Amplifiers

In the previous lecture, it was stated that transistors are often used to create amplifiers. An **amplifier** is a device that increases the power of a weak electronic signal.

15.1 Gain and Decibels

An amplifier is typically represented in a circuit drawing as an isosceles triangle like the one shown in Fig. 15.1. The two DC voltages V_{cc+} and V_{cc-} are known as the "voltage rails", and they supply the electrical power that is given to the signal. The amplifier takes the input signal $V_{in}(t)$ and scales it by a constant factor G, such that

$$V_{out}(t) = GV_{in}(t), \tag{15.1}$$

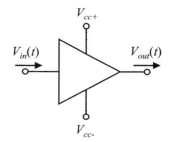

Figure 15.1: A generic amplifier adds power to a signal $V_{in}(t)$.

where $G = V_{out}/V_{in}$ is formally known as the amplifier **gain**. Recall that the instantaneous electrical power in an AC circuit is the product of current and voltage $\dot{Q}(t) = I(t)V(t)$, so an amplifier may also increase the power by increasing the amplitude of the current.

Amplification can also be quantified by a logarithmic scale known as **decibels (dB)**. A decibel of amplification is defined by the equation

$$1dB = 10\log_{10}\left(\frac{|\dot{Q}_{out}|}{|\dot{Q}_{in}|}\right), \tag{15.2}$$

Notes: The output is limited by the voltage rails, such that $V_{cc-} < V_{out} < V_{cc+}$.

where $|\dot{Q}_{out}|$ and $|\dot{Q}_{in}|$ are the amplitudes of the output and input power, respectively. For a resistive load, the power $|\dot{Q}|$ is proportional to $|V|^2$, so Eq. (15.1) can be rewritten as

$$1dB = 20\log_{10}\left(\frac{|V_{out}|}{|V_{in}|}\right), \tag{15.3}$$

where a log identity has allowed us to pull the exponent of 2 out in front, and any constants of proportionality have cancelled each other out.

15.2 Signal-to-Noise Ratio

The most common reason for using an amplifier is to increase the **signal-to-noise ratio (SNR)**. Imagine you are speaking to your mother on the phone at a tailgate party. It is difficult hear her voice over the background noise of the party, so you turn up the volume on your phone. Doing so, you have effectively increased the signal-to-noise ratio of your mother's voice relative to the noise of the party.

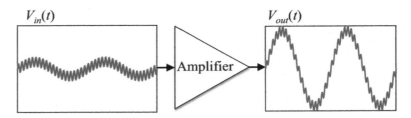

Figure 15.2: An amplifier is used to increase the signal-to-noise ratio.

The same concept can be generalized to any measurable signal $V(t)$. Consider the low frequency signal shown in Fig. 15.2 with a high frequency noise component. In the previous chapter, we discussed using a low-pass filter to remove the noise. Another approach is to amplify the signal $V(t)$ relative to the noise, thus increasing the signal-to-noise ratio.

15.3 Differential Amplifiers

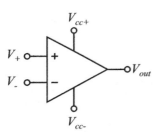

Figure 15.3: A differential amplifier has two inputs and one output.

Looking at Fig. 15.2 might leave you wondering how the signal can be amplified without amplifying the noise. This is often achieved using a **differential amplifier**. The differential amplifier, shown in Fig. 15.3, has two inputs and one output

$$V_{out} = G_d(V_+ - V_-), \tag{15.4}$$

where V_- is known as the inverting input, V_+ is known as the non-inverting input, and G_d is the differential gain or "open-loop" gain. For most amplifiers, the open-loop gain is $G_d > 10^4$.

When used correctly, a differential amplifier can effectively cancel out noise using the two inputs. The amount of noise cancellation that a differential amplifier is capable of achieving is quantified by the **common mode rejection ratio (CMRR)**. A large value of CMRR implies that the amplifier is capable of cancelling out a large amount of noise.

15.4 Op-Amps

Operational amplifiers or **op-amps** are small, inexpensive differential amplifiers sold as small IC chips. Op-amps are typically connected in circuits that utilize **feedback** from the output.

A simple op-amp circuit that utilizes feedback is shown in Fig. 15.4. This circuit—known as a **follower**—has the output fed back directly to the non-inverting input, such that $V_+ = V_{out}$. The input is connected directly to the inverting input, so $V_- = V_{in}$. Substituting into Eq. (15.4) gives us

$$V_{out} = G_d(V_{out} - V_{in}), \tag{15.5}$$

which can be rearranged to be

$$V_{out} - V_{out}(t)/G_d = V_{in}. \tag{15.6}$$

Because the open-loop gain $G_d > 10^4$, we may neglect the second term, leaving us with the simple relation that $V_{out} = V_{in}$ for the follower circuit.

At a glance, it seems the follower circuit is not very useful. However, recall that electrical power is the product of *both* voltage *and* current. The follower circuit actually amplifies current $I(t)$, while leaving the voltage $V(t)$ unchanged. This is useful for preventing the voltage droop discussed in Section 5.4.

Notes: A typical value for the CMRR for an op-amp is around 80 dB.

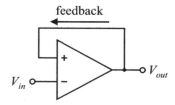

Figure 15.4: A simple follower circuit uses feedback.

15.5 The Golden Rules of Op-Amps

Feedback can be confusing to even the most savvy engineering student. Luckily, there are two simple rules that make it easy to understand and analyze any op-amp circuit with feedback.

1. The inputs V_+ and V_- draw approximately zero current into the amplifier, while the amplifier output V_{out} is capable of providing a large amount of current.

2. The output voltage V_{out} will change itself to keep the inputs at the same voltage, such that $V_+ = V_-$.

The following example demonstrates how these rules are used to analyze an op-amp circuit.

Notes: It is common to omit the voltage rails V_{cc+} and V_{cc-} from amplifiers circuit drawings.

Example 15.1:

Derive a formula for the closed-loop gain G in terms of the resistances R_1 and R_2 for the Op-Amp circuit shown below.

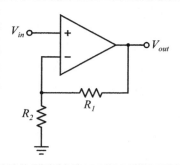

Solution: First, we sketch the current flow from the output through the two resistors R_1 and R_2.

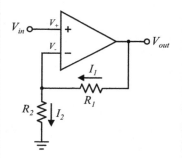

According to Rule 1, no current flows into the amplifier, so $I_1 = I_2 = I$. Using Ohm's law to calculate the voltage drops across the resistors gives us $V_- = IR_2$, and $V_{out} = I(R_1 + R_2)$.

Dividing these equations yields $\frac{V_-}{V_{out}} = \frac{R_2}{R_1 + R_2}$.

Next, we apply Rule 2, which says that $V_+ = V_-$. The input V_{in} is connected directly to the non-inverting input, so $V_{in} = V_+ = V_-$. Combining this with the previous equation obtained from Rule 1, we have $\frac{V_{in}}{V_{out}} = \frac{R_2}{R_1 + R_2}$.

Inverting this gives us the closed-loop gain, $\frac{V_{out}}{V_{in}} = \frac{R_1 + R_2}{R_2}$.

This can be algebraically simplified to $\boxed{G = 1 + \dfrac{R_1}{R_2}}$.

15.6 More Op-Amp Circuits

Op-amps can also be used to mathematically manipulate a signal $V(t)$. For example, the output of the **integrator** circuit shown in Fig. 15.5 is proportional to the integral of the input signal, such that

$$V_{out}(t) = \frac{1}{RC} \int V_{in}(t)\, dt. \tag{15.7}$$

Similarly, the output of the **differentiator** circuit is proportional to the derivative of the input signal, such that

$$V_{out}(t) = -RC\frac{dV_{in}}{dt}. \tag{15.8}$$

The simple RC filters we saw in Lecture 13 were *passive* filters, meaning that they did not increase the power of the signal. Circuits similar to the integrator and differentiator can be used to make low-pass and high-pass *active* filters, which both remove unwanted oscillations and increase the signal-to-noise ratio.

Exercises 15:

1. An amplifier receives an input signal given by the formula $V_{in}(t) = (100\text{mV})\sin[(500\text{rad/s})t]$, and it outputs a signal given by $V_{out}(t) = (10\text{V})\sin[(500\text{rad/s})t]$. Calculate the gain in decibels (dB).

2. Consider the following scenarios, and identify what is the "signal" and what is the background "noise".

   (a) A football player cannot hear his coach, because the crowd is cheering too loudly.

   (b) A dense fog settles in over an airport, and a pilot cannot see the lights on the runway.

   (c) An astronomer cannot see the stars during the day.

3. Use the golden rules of op-amps to derive a formula for the closed-loop gain G in terms of the resistances R_1 and R_2 for the Op-Amp circuit shown below.

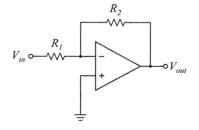

Integrator

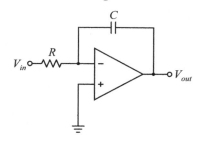

Differentiator

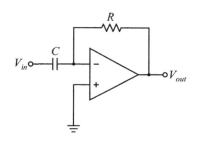

Figure 15.5: Op-amps can be used to integrate (top) or differentiate (bottom) a signal.

Notes: Op-amp circuits can be considered **analog computers,** if they perform mathematical operations on an analog signal.

16 Digital Electronics

An analog sensor outputs a *single* voltage that is mathematically related to some physical parameter, such as temperature or pressure. In a **digital computer**, a physical parameter is represented by *many* voltages that form a **binary code**. In this lecture, we will discuss the fundamentals of digital electronics.

16.1 Boolean Logic

Transistors can be either ON or OFF, depending on the gate voltage, and multiple transistors can be connected together to create a logic circuit. **Boolean logic** is the mathematical formalism used to understand systems where the variables have two possible states: ON or OFF, True or False, 1 or 0, etc. For example, an engineer might want to shut off a hot water heater if the water is above a certain temperature, or if the pressure is above a certain threshold. These variables all have two possible states that are either True or False: The water is too hot? The pressure is too high? The water heater is ON?

Boolean variables are fed into a network of **logic gates** that perform various logical operations. Shown below in Fig. 16.1 are the three fundamental logic gates along with their **truth tables**.

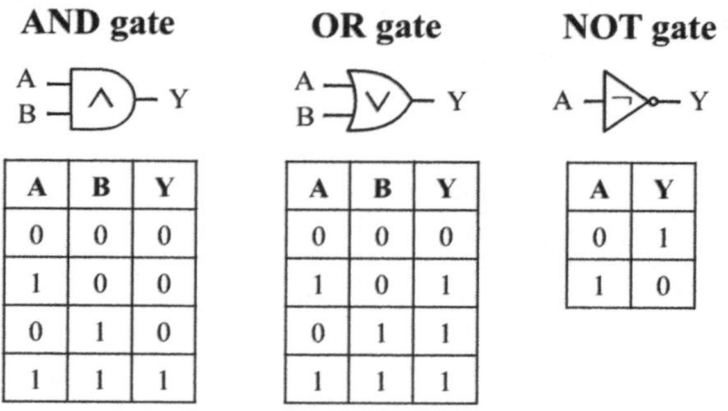

AND gate			**OR gate**			**NOT gate**	
A	**B**	**Y**	**A**	**B**	**Y**	**A**	**Y**
0	0	0	0	0	0	0	1
1	0	0	1	0	1	1	0
0	1	0	0	1	1		
1	1	1	1	1	1		

Figure 16.1: The three fundamental Boolean logic gates and their truth tables.

- The **AND gate** outputs 1 if and only if both A *and* B are 1. Algebraically, this is written as $Y = (A \wedge B)$.

- The **OR gate** outputs 1 if A *or* B is 1. Algebraically, this is written as $Y = (A \vee B)$.

- The **NOT gate** simply inverts a 0 to a 1, and vice versa. Algebraically, this is written as $Y = \neg A$.

Example 16.1:
Write out the truth table for the array of logic gates shown below.

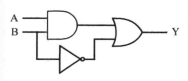

Solution: The logic represented by the gate array can be written algebraically as $(A \wedge B) \vee (\neg B)$. Performing each operation from left-to-right generates the truth table.

A	B	$A \wedge B$	$\neg B$	$(\mathbf{A} \wedge \mathbf{B}) \vee (\neg \mathbf{B})$
0	0	0	1	1
0	1	0	0	0
1	0	0	1	1
1	1	1	0	1

16.2 Binary Code

We typically use a base-10 number system, where we roll over to a new digit after cycling through 10 numbers. There is nothing special about base-10 math, and we only use it because we have 10 fingers on our hands. We could just as easily roll over to a new digit after counting to 8 or 6 or any number.

Binary code is a base-2 number system, where numbers are represented by a string of "ones and zeros". For example, the number 51 in 8-bit binary code is 00110011. This is how numbers are represented in a digital computer. Transistors are either ON or OFF, depending on the gate voltage. If the threshold voltage for the transistors is 3V, then 3 Volts represents a 1, and 0 Volts represents 0. Therefore, multiple voltages on multiple transistors can be combined to represent an integer number in binary code. Furthermore, transistors can be connected in circuits to create the Boolean logic gates in Fig. 16.1. Logic gates constructed from transistors can be further combined to perform basic mathematical operations, such as binary addition. This is the basis of the modern digital computer.

Example 16.2:

Show that the 8-bit binary number 00110011 is equal to 51.

Solution: Instead of a ones place (10^0), a tens place (10^1), a hundreds place (10^2), etc., binary code has a ones place (2^0), a twos place (2^1), a fours place (2^2), etc. This can be illustrated by the following table.

bit value	2^7	2^6	2^5	2^4	2^3	2^2	2^1	2^0
bits	0	0	1	1	0	0	1	1
voltages	0V	0V	3V	3V	0V	0V	3V	3V

Adding the numbers from left-to-right, we have
$$0(2^7) + 0(2^6) + 1(2^5) + 1(2^4) + 0(2^3) + 0(2^2) + 1(2^1) + 1(2^0) = 51$$

16.3 Single and Double Precision Floating Point Variables

In Example 16.1, we looked at the simple integer number 51. To represent a rational number in scientific notation, most computers use a standard format known as a **floating point variable**. The structure of floating point variables is illustrated in Fig. 16.2. A single precision "float" is composed of 32 bits (4 bytes), and a double precision float is composed of 64 bits (8 bits).

Single precision floating-point variable

1 bit	8 bits	23 bits
sign	exponent	fraction

Double precision floating-point variable

1 bit	11 bits	52 bits
sign	exponent	fraction

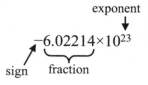

Figure 16.2: Rational numbers are typically represented in a computer using the floating point data structures illustrated here.

16.4   Parallel vs. Serial Communication

Our modern world is now dominated by digital communication, and it is used in nearly every aspect of life, from banking to romance. At the most fundamental level, digital data is sent as either voltage pulses through a wire, pulses of light through a fiber optic cable, or pulses of a radio wave from an antenna. These pulses represent the bits of a binary code.

As shown in Fig. 16.3, bits of information can be transmitted via parallel or serial communication protocols. In both types of data transmissions, the communicating computers use pulses from an internal clock to determine where one bit ends and a new bit begins.

- **Parallel Communicationparallel communication** – Multiple wires are used to transmit the bits simultaneously.

- **Serial Communicationserial communication** – A single wire is used to transmit all bits. Digital files are typically sub-divided into "packets" of a certain time and bit length.

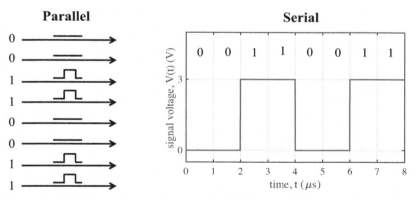

Figure 16.3: An 8-bit number can be transmitted (left) in parallel on multiple wires or (right) as a serial packet on a single wire.

As computers have become faster, serial protocols have become the dominant form of digital communication. Familiar forms of serial data transmission include the Universal Serial Bus (USB) and the Internet Protocol (IPv4). The USB 3.0 protocol divides files into 1024 byte packets, while IPv4 divides files into 65,535 byte packets.

Notes: For USB, the **baud rate** refers to the speed that data is transmitted in bits per second (bps). A baud rate of 9600 bps is the default for most devices.

Exercises 16:

1. Consider logic gate array shown below, which is known as a **half adder**.

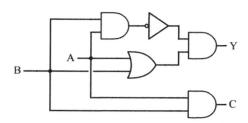

(a) Write an algebraic formula for C in terms of A and B.

(b) Write an algebraic formula for Y in terms of A and B.

(c) Write out the truth table (similar to Example 16.1), and show that the 2-bit string formed by the two outputs CY is equal to the sum of the input bits $A + B$.

R   G   B

2. Most electronic displays use red, green, and blue (RGB) pixel arrays. That is, a single pixel actually contains three sub-pixels, and the intensity or "brightness" of each sub-pixel can be set by an 8-bit integer value between 0 and 255. Different combinations of RGB sub-pixel intensities can be used to create nearly any color in the visible spectrum.

(a) How many different discrete colors can be created using all possible combinations of the 3 sub-pixel intensities?

(b) Consider a 16 megapixel image composed of RBG pixels. How many distinct images are possible? (Express your answer as a power of 2. i.e., 2^N)

3. A 40 million byte video file is sent over the internet via IPv4. Approximately how many serial packets will be created to transmit the entire file?

17 Digital Sensors

Digital sensors typically output a digital pulse train that is indicative of some measured quantity. These pulses can represent discrete events, such as the detection of a photon, or they can represent serial binary code packets.

17.1 Counters and Timers

There are a variety of sensors and detectors that measure discrete events. For example, the spoke of a wheel passes by an optical sensor. This can be considered a binary or Boolean measurement, if there are only two possibilities: the wheel spoke is near the optical sensor, or it is not. This is represented as a voltage pulse from the sensor that is either HIGH or LOW.

A **digital counter** is an integrated circuit (IC) chip that simply counts the number of times a voltage signal changes from LOW to HIGH (called a "rising edge trigger") or HIGH to LOW (called a "falling edge trigger"). The number of counts is represented as a digital binary code within the counter IC chip.

There are many different counter chips on the market, and almost all have the following features.

- **Clock (CLK) input** - The counter increments the count when either a rising edge or falling edge trigger (depending on the chip) is detected by the clock input.

- **Enable (ENBL) input** - The counter will only change the count if the enable input pin is set to HIGH (or LOW depending on the chip).

- **Reset (RST) input** - The count is reset to 0 when the reset input pin is set to HIGH (or LOW depending on the chip).

- **Digital output (Q0, Q1, Q2, ...)** – The count is represented as a binary code on several output pins with the bits in parallel, where Q0, Q1, Q2, ... each represent one of the bits.

Notes:

- A **rising edge trigger** increments when the signal transitions from LOW to HIGH.

- A **falling edge trigger** increments when the signal transitions from HIGH to LOW.

A **timer circuit** simply adds a periodic clock signal to the measurement that keeps track of the time. The **clock signal** is just a simple square wave that oscillates at a precise frequency. For example, a 1 kHz clock signal connected to a counter would create a timer that measures the elapsed time in milliseconds.

Timers are often used for measuring the time that passes between two events, where signals from external sensors are used to start, stop, and reset the counter. Complex timing sequences involving multiple **trigger signals** can be created by connecting logic gates to the enable pin of the counter. These timing sequences are often explained using a **timing diagram**, such as the one shown in Fig. 17.1, where Trigger A starts the counter, and Trigger B stops the counter.

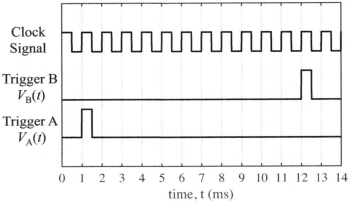

Figure 17.1: A timing diagram for a timer circuit, where Trigger A starts the count and Trigger B stops the count.

The following example illustrates a practical use for this type of timing system.

Example 17.1:
Recall that Galileo's famous inclined plane experiment measures the time t it takes the ball to roll a distance x down an inclined plane. A modern version of Galileo's inclined plane experiment uses two photogates connected to a counter-timer. A photogate consists of an LED and a photoresistor circuit, which outputs a LOW voltage when light from the LED is incident on the photoresistor. When the ball blocks the light, the photogate outputs a HIGH voltage.

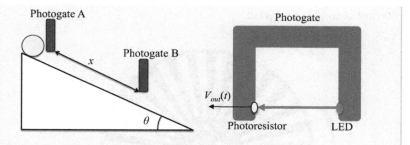

When the ball passes through Photogate A, it briefly outputs a HIGH voltage. This signal is fed into a counter/timer circuit as Trigger A, and its rising edge of the signal triggers the timer to *start* counting.

When the ball passes through Photogate B, it briefly outputs a HIGH voltage. This signal then triggers the timer to *stop* counting.

This timing sequence is illustrated by the timing diagram in Fig. 17.1.

17.2   Digital Tachometer

A **digital tachometer** is a device that measures the angular speed of a spinning mechanical part. Either an optical or magnetic sensor is used to detect the passage of some reflective tape or a small, magnetized bead or ferromagnetic protrusion (see Fig. 8.3). The output of the sensor is used as a trigger to both start and stop a timer circuit. The angular speed in RPMs is then inversely proportional to the digital time count between two successive triggers.

17.3   Quadrature Encoder

A **quadrature encoder** is a sensor that measures the angular position of a rotating mechanical part. The device consists of a slotted wheel, mounted onto the rotating shaft or axle, with two photogates that detect the passage of the slots on the wheel, as illustrated in Fig. 17.2. The output of the photogates is connected to a counter circuit that counts the pulses. The pulse count is converted to an angle of rotation based on the number slots, or counts per revolution (CPR), on the encoder wheel.

Again, note that the encoder has *two* photogates. This allows the counter circuit to determine the *direction* of rotation based on the phase between the two photogate signals A and B. Looking at Fig. 17.2, we see that the output from photogate B will *lead* A if the wheel is rotating clockwise. Conversely, the output from B will *lag* behind A if it is rotating counter-clockwise.

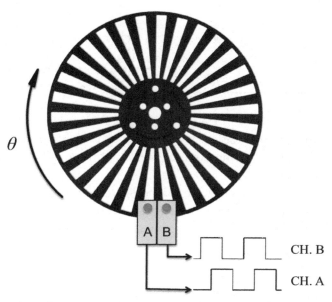

Figure 17.2: A quadrature encoder consists of a slotted wheel mounted on a rotating shaft. The photogates detect the slots in the wheel as the shaft is rotated.

17.4   Digital Sensor Output: I²C, SPI, UART, and CAN

There are a variety of electronic sensors that output a digital signal. Many of these are essentially analog sensors, but with an additional IC chip that converts the analog voltage to a digital value using a calibration formula. This has three main benefits. First, it eliminates the step of calibrating the analog sensor and converting the voltage output using the calibration formula. Second, the digital output can be fed directly into a computer or microcontroller (more on microcontrollers in the next lecture). Third, it allows dozens of different sensors to be connected to a computer or microcontroller using only two wires. That is, the sensors all send their data through a single set of shared wires, known as a **data bus**, as illustrated in Fig. 17.3

Only one device may send data on the bus at any given time, and hardware can be damaged if multiple devices try to send their data at the same time. To prevent this, one device is designated the "master", while the other devices are designated as "slaves". The master device determines which slave device gets to transmit data on the bus at any given time.

Listed below are some common serial communication protocols for digital communication between sensors and other devices.

Notes: The "master/slave" nomenclature in an unfortunate relic that has been around for several decades. There has been a recent push to revise it to something less barbaric. Can you think of an alternative?

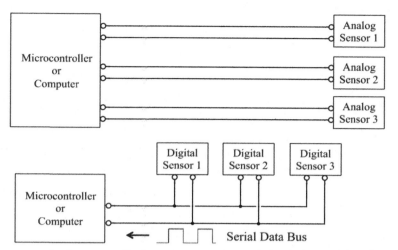

Figure 17.3: Analog sensors (top) require more wires and electrical connections than digital sensors connected to a serial data bus (bottom).

Notes: Manufacturing costs and the probability of failure both increase with the number of wires in a product. Thus, the advantage of digital sensors is that they can all share a single set of wires known as a **data bus**.

- **UART (Universal Asynchronous Receiver-Transmitter)** is one of the earliest serial communication protocols. It is "asynchronous", which means that there is no shared clock signal to determine where one bit starts and another bit ends (see Fig. 16.3). Thus, sensors with UART output would need to have their own internal clock, which adds to the cost.

- **SPI (Serial Peripheral Interface)** uses a bus with four wires: two wires for sending data (MOSI and MISO), a wire for the clock signal (SCLK), and a wire known as the "slave select" (SS) that determines which device is allowed to transmit data. It is the basic protocol used by SD memory cards.

- **I^{2}C (pronounced "eye square see")** is popular, because the bus contains only two wired connections: a wire for the data (SDA) and a wire for the clock signal (SCL).

- **CAN (Control Area Network)** is a standard data bus configuration for digital sensors and microcontrollers used in motor vehicles such as cars, trucks, and tractors.

Exercises 17:

1. A timer is created by connecting a 10 kHz square wave to the clock input of an 8-bit counter.

   (a) What is the sensitivity or "resolution" of the timer? That is, what is the smallest time increment it can measure in milliseconds?

   (b) The 8-bit output of the counter can represent integers from 0 – 255 (00000000 to 11111111 in binary). Based on this, what is the range of the timer in milliseconds?

   (c) The clock signal is changed to a lower frequency square wave at 100 Hz. Calculate the new sensitivity and range.

   (d) Is there a trade-off between sensitivity and range for the timer? Explain.

2. Search either the Digikey or Mouser website for a quadrature encoder that mounts to a 6 mm shaft.

   (a) Write down its part number, price is US dollars, and counts per revolution.

   (b) Look in the data sheet for a CAD drawing of the encoder. Print out the page with the drawing.

3. Search either the Digikey or Mouser website for a pressure sensor with an I^2C digital output.

   (a) Print out the data sheet.

   (b) Write down the part number, price is US dollars, and pressure range.

18 Analog-to-Digital Conversion

An **analog-to-digital converter (A/D)** is an electronic device that measures an analog voltage, converts it to a digital value (represented in binary code), and saves it in computer memory. Thus, a continuous analog signal $V(t)$ becomes a set of digital numbers recorded periodically at discrete moments in time.

A **digital-to-analog converter (D/A)** is an electronics device that takes a digital value from computer memory and outputs it as a voltage.

Notes: Analog-to-digital and digital-to-analog converters are often referred to collectively as **I/O streams** (input/output).

18.1 Discrete Digital Signals

An analog signal $V(t)$ can be digitally recorded into computer memory using an A/D. Illustrated in Fig. 18.1, the continuous analog signal $V(t)$ is converted to a set of discrete voltages $[V_1, V_2, V_3, \ldots V_N]$, each measured at a specific moment in time $[t_1, t_2, t_3, \ldots t_N]$. The time between samples and overall quality of the recorded digital signal are dictated by the **sampling frequency** f_s. There are a few important pa-

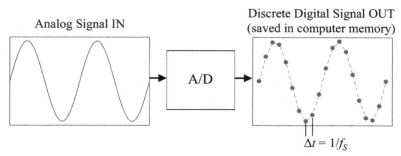

Figure 18.1: An analog-to-digital converter (A/D) measures an analog voltage and converts it to a discrete digital value that is saved in memory.

rameters that you must understand when using an A/D to record a signal.

- **Sampling frequency** f_s determines the rate at which data points are recorded. It may be specified in units of Hz or samples/second. Higher sampling frequencies increase the quality of the discrete digital signal and make it smoother.

- **The time between samples** Δt is determined by the sampling frequency, such that $\Delta t = \frac{1}{f_s}$.

- **The duration** T_{max} is the length of time over which data is recorded.

- **The number of data points collected** N is determined by the duration and the sampling frequency and the duration. Higher sampling frequencies and longer durations will result in a large number of data points.

The duration, number of data points, and sampling frequency are all related by the formula

$$T_{max} = \frac{N-1}{f_s}. \tag{18.1}$$

Analog-to-digital converters may not automatically record a set of time values for the measured voltages. If this is the case, you must generate the time data yourself using the sampling frequency. Each discrete time value corresponds to an integer multiple of Δt, such that

$$[t_1, t_2, t_3, \ldots t_N] = [1, 2, 3, \ldots N]/f_s. \tag{18.2}$$

18.2   *The Nyquist Frequency*

The quality of a digital signal depends on the sampling frequency. If the sampling frequency is too low, high frequency components of a signal will not be resolved. Consider the continuous analog signal shown in Fig. 18.2, which contains both a high frequency and a low frequency component. The discrete digital signal recorded by the A/D does not resolve the high frequency component, because the sampling frequency is too low.

According to the Nyquist Theorem, a digital signal recorded with an A/D cannot resolve frequencies greater than half the sampling frequency. That is, the maximum frequency that can be resolved is known as the **Nyquist frequency**

$$f_{max} = \frac{f_s}{2}. \tag{18.3}$$

Another way to put it is that you must have at least two samples per cycle of a sine wave.

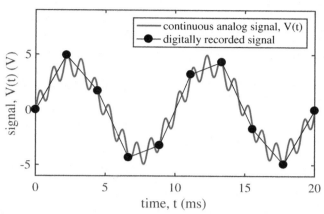

Figure 18.2: The sampling frequency is too slow to resolve the high frequency component of the analog signal.

This brings about another interesting engineering trade-off involving the sampling frequency:

- High sampling frequencies can resolve fast oscillations, but results in very large data sets that can overload a computer's memory.

- Low sampling frequencies result in small data sets that are easy for a computer handle, but cannot capture any fast oscillations in a signal.

Example 18.1:

A 16-bit A/D is used to collect data at a rate $f_s = 25,000$ Samples/s for 3 minutes, and the data is transferred to a computer.

- What is the maximum frequency component of the signal that can be resolved?

Solution: $f_{max} = \frac{f_s}{2} = 12,500$ Hz

- How many data points are collected?

Solution: $N = \left(\frac{25000 \text{ samples}}{\text{second}} \right) (3 \text{ minutes}) \left(\frac{60 \text{ seconds}}{\text{minute}} \right)$
$= 4.5 \times 10^6$ samples

- The data points are saved to the computer as double-precision floating-point variables. Nominally, how much computer memory will the data use?

Solution: $(4.5 \times 10^6 \text{ samples}) \left(\frac{8 \text{ bytes}}{\text{sample}} \right) = 36 \times 10^6$ bytes

> • A student wishes to use Matlab to analyze the data. The computer has allocated 510 million bytes of memory to Matlab. At the given sampling rate, how long can data be collected for before Matlab runs out of memory and crashes?
>
> *Solution:* $t = 510 \times 10^6$ bytes $\left(\frac{1 \text{ sample}}{8 \text{ bytes}} \right) \left(\frac{1 \text{ second}}{25000 \text{ samples}} \right)$
> $= 2550$ seconds $= 42$ minutes

Notes: Microcontrollers essentially form the "brains" of many modern consumer products, such as washing machines and cars. PLCs are primarily used to automate factory equipment.

18.3   Microcontrollers and PLCs

A **microcontroller** is a rudimentary computer packed into a small IC that can be programmed to automate various tasks. Most have a few A/D inputs that allow them to record signals from analog sensors, as well as D/A outputs that allow them to control various actuators. Unlike a personal computer, a microcontroller does not have an operating system. Rather, they run pre-programmed code sequences that always operate as an infinite loop. That is, a microcontroller will repeatedly run the same set of code instructions over and over again.

A **programmable logic controller (PLC)** is a small, rugged computer that is used to automate various processes in a factory setting. For example, a PLC can be used to turn off a heater when a vat of liquid reaches a certain temperature, or it might synchronize a conveyor belt with various pneumatic actuators. PLCs are usually "modular", meaning that multiple PLCs and various peripherals can be connected to increase the number of digital or analog I/O streams.

Notes: Microcontrollers are typically programmed using a C programming language. PLCs are usually programmed using a *graphical* programming language known as "ladder logic".

18.4   Pulse Width Modulation (PWM)

Digital devices such as microcontrollers often approximate a DC voltage output using **pulse width modulation (PWM)**. Shown in Fig. 18.3, a PWM signal is merely a square wave oscillating between HIGH and LOW states. The **duty cycle** is the percentage of the period that the square wave is in the HIGH state. The average voltage of the signal is simply the HIGH voltage (typically 5V) multiplied by the duty cycle as a fraction

$$\langle V \rangle = \left(\frac{\%\text{DutyCycle}}{100\%} \right) V_{HIGH}. \tag{18.4}$$

The PWM output from a microcontroller or computer is usually very weak. It is common to amplify a PWM signal using a single transistor, as shown in the following example.

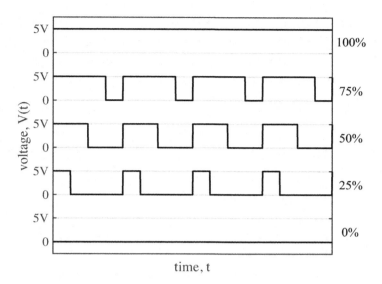

Figure 18.3: PWM signals are plotted with various duty cycles ranging from 0 to 100%.

Example 18.2:

An engineering student wants to use an Arduino UNO micro-controller to control the speed of a 12V DC electric motor using its PWM output. Unfortunately, the PWM signal is not powerful enough (does not output enough current) to drive the motor.

Solution: The PWM signal is not powerful enough to directly drive the motor, but it is powerful enough to drive the gate of a MOSFET transistor. (Recall that transistors are commonly used to amplify weak signals.) Shown below, the motor is connected to a 12V DC power supply and connected in series with the source and drain of the transistor.

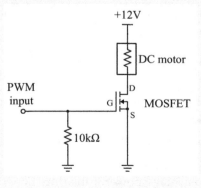

Current flows from the 12V DC power supply through the motor when the PWM signal drives the gate HIGH. The average current through the motor increases with the average value, or duty cycle, of the PWM signal, thus causing the motor to spin faster. The microcontroller is able to adjust the PWM duty cycle, and thus control the motor speed.

Exercises 18:

1. A 16-bit A/D is used to collect data at a rate $f_s = 50,000$ Samples/s for 2 minutes, and the data is transferred to a computer.

   (a) How many data points are collected?

   (b) The data points are saved to the computer as double-precision floating-point variables. Nominally, how much computer memory will the data use?

   (c) What is the maximum frequency component of the signal that can be resolved?

   (d) A student wishes to use Matlab to analyze the data. The computer has allocated 610 million bytes of memory to Matlab. At the given sampling rate, how long can data be collected for before Matlab runs out of memory and crashes?

2. Search the National Instruments website (NI.com) for an I/O device with at least 4 analog input channels that connects to a computer via USB.

   (a) Write down its model number, price is US dollars, the maximum sampling frequency

   (b) Download the data sheet for the device. Print out the page with the drawing of the device "pinout" that shows where the various electrical connections are made.

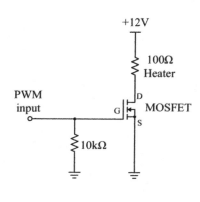

3. Consider the adjacent circuit, where a weak 500Hz PWM signal is used to switch the transistor ON and OFF.

   (a) Sketch a plot of voltage across the resistor vs. time for a 30% duty cycle.

   (b) Sketch a plot of power dissipated in the resistor vs. time with a 30% duty cycle.

   (c) Derive and equation of the average power dissipated as a function of the % duty cycle.

   (d) Calculate the average power if the duty cycle is 80%.

19 Discrete Digital Signal Processing

The next four lectures will present several mathematical tools for processing and understanding measured transient signals. We will begin with a few basic techniques used for the type of discrete digital signals introduced in the previous lecture.

19.1 Moving Average and Moving Standard Deviation

The **moving average** is typically used to smooth a noisy signal. It essentially replaces each data point V_i with the average of the adjacent data points $\overline{V}_{Mi}$. A central moving average is typically used in science and engineering, where the average of m data points to the left and m to the right are taken. A moving average with a width of $m = 1$ is illustrated in Fig. 19.1. The moving average at each data point is given by the formula

$$\overline{V}_{Mi} = \sum_{k=i-m}^{i+m} \frac{V_k}{2m+1},$$ (19.1)

where m determines the number of points used in the moving average. Large values of m will result in more smoothing of the data. Note that the moving average cannot be computed for the first m data points and last m data points, as the index of the summation would go "out of bounds". (This is illustrated by the gray cells in Fig. 19.1.) Importantly, the moving average—or any type of smoothing filter—will decrease the amount of information contained in a data set.

Notes: In finance, it is common to use a *left* moving average and standard deviation, which averages over m data points to the left, but not the right.

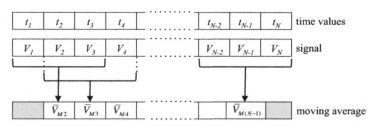

Figure 19.1: The moving average iterates through the data and calculates the average of each "bunch" of data. The diameter of each bunch shown above is $m = 1$.

The **moving standard deviation** quantifies how noisy or volatile a signal is at a certain time. In finance and investment, it is often used to quantify the "volatility" of a stock price. The moving standard deviation is calculated similar to the moving average by taking the standard deviation of m data points to the left and m to the right,

$$s_{Mi} = \sqrt{\sum_{k=i-m}^{i+m} \frac{(V_k - \overline{V}_{Mi})^2}{2m+1}}, \tag{19.2}$$

Note that the formula for the moving standard deviation contains the moving average $\overline{V}_{Mi}$.

Example 19.1:

Write a Matlab script that computes and plots the moving average over $m = 30$ days in both directions for the daily NASDAQ composite plotted below. The data has been downloaded from the internet as a .CSV (comma separated value) file.

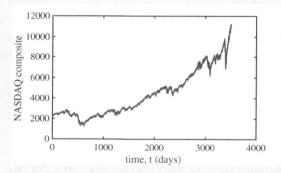

Solution: The Matlab script reads in the daily NASDAQ values from the CSV file, and stores them into a vector V. In line 15 - 17, a for loop iterates over the vector V and takes the average of each "bunch" of data.

A subset of data from a vector in Matlab can be extracted via the syntax $V(a : b)$, where a and b are the indices of the beginning and end of the subset, respectively. For example, $V(6 : 20)$ will output a new vector containing the elements from $i = 6$ to $i = 20$.

The resultant plot should be much smoother than the original, as the high frequency noise has been "averaged out" of the signal. Note that m data points at the beginning and end will be arbitrarily set to zero, to keep the for loop from going out of bounds on the data set.

Matlab Script:

```
1 -    clc
2 -    close all
3
4      %read in NASDAQ data
5 -    V = csvread('NASDAQ_data.csv')';
6
7 -    N=length(V); %get the number of data points
8 -    m=30; %width of moving average
9
10 -   t=linspace(1,N,N);  %create a time vector, units of days
11
12 -   Vm=zeros(N,1); %intialize column vector for moving average
13
14     %Use a for loop to iterate over the data
15 -   for n=m+1:N-m
16 -        Vm(n)=mean(V(n-m:n+m)); %average over the range from n-m to n+m
17 -   end
18
19     %Raw data
20 -   figure(1)
21 -   plot(t,V,'lineWidth',2)
22 -   xlabel('time, t (days)')
23 -   ylabel('NASDAQ composite')
24     %Make fonts big, set aspect ratio to 1.61
25 -   set(gca,'FontSize',16,'linewidth',1.5,'FontName',...
26         'Times New Roman')
27 -   set(gcf, 'Position',  [100, 100, 450, 450/1.61])
28
29
30     %Filtered with moving average
31 -   figure(2)
32 -   plot(t,Vm,'lineWidth',2)
33 -   xlabel('time, t (days)')
34 -   ylabel('NASDAQ (30 day moving avg.)')
35     %Make fonts big, set aspect ratio to 1.61
36 -   set(gca,'FontSize',16,'linewidth',1.5,'FontName',...
37         'Times New Roman')
38 -   set(gcf, 'Position',  [100, 100, 450, 450/1.61])
```

Resultant Plot:

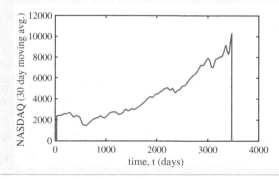

19.2   Numeric Differentiation

The derivative is defined as the slope of the tangent line at any given point on the curve. For a discrete digital data set, this can be estimated by the "rise-over-run" between the two data points, as illustrated in Fig. 19.2. Dividing the "rise" ΔV by the "run" Δt gives us the **finite difference** formula

$$V'(t_i) = \frac{V_i - V_{i-1}}{t_i - t_{i-1}} \tag{19.3}$$

where i is the integer index of the ith data point and $V'(t_i)$ is the numeric derivative at the time t_i. (Note that the calculation will have to begin with the 2nd data point, lest the $i-1$ index go out of bounds.)

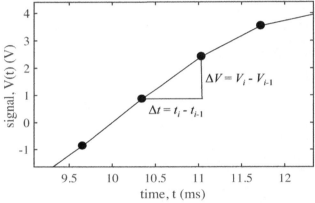

Figure 19.2: The derivative approximated by the slope of the line connecting two adjacent data points.

Notes: Recall that Δt is determined by the sampling frequency $\Delta t = 1/f_s$.

19.3   Numeric Integration

The integral of a function is essentially the "area under the curve " over a certain range. The area under the curve can be calculated by drawing rectangles with their base at $V = 0$ and one of their opposite corners touching each data point V_i, as illustrated in Fig. 19.3. This method of numeric integration is known as a **Riemann sum**. This Riemann sum adds up the area of each individual rectangle via the formula

$$\int_a^b V(t)dt \approx \sum_{i=1}^{N} V_i \Delta t, \tag{19.4}$$

where a and b are the limits of integration with $t_1 = a$ and $t_N = b$.

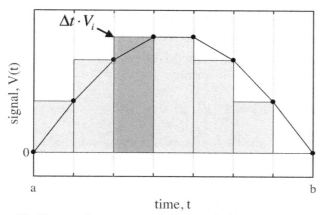

Figure 19.3: The Riemann Sum computes the area under the curve by summing the area of each rectangle $\Delta t V_i$.

A more accurate method for calculating the area under the curve uses trapezoids, rather than squares. This method of numeric integration, shown in Fig. 19.4, is known as the **trapezoidal method**. Thus, the area under the curve is computed by summing the area of each individual trapezoid via the formula

$$\int_a^b V(t)dt \approx \sum_{i=1}^{N} \frac{V_{i-1} + V_i}{2} \Delta t. \tag{19.5}$$

Writing out the terms in the summation gives us

$$\int_a^b V(t)dt \approx \frac{\Delta t}{2}[V_0 + 2V_1 + 2V_2 + \ldots + 2V_{N-1} + V_N], \tag{19.6}$$

which is commonly referred to as the **trapezoidal rule**.

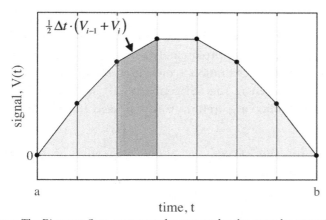

Figure 19.4: The Riemann Sum computes the area under the curve by summing the area of each trapezoid $\frac{1}{2}\Delta t(V_{i-1} + V_i)$.

It should also be noted that the equations listed above yield a single value for the **definite integral**, which is the area under the curve from a to b. If we wish to compute the **anti-derivative** $F(t_i)$, we must compute the integral for every single data point at t_i. That is, the anti-derivative can be calculated using the formula

$$F(t_i) = \int_0^{t_i} V(t)dt \approx \sum_{k=1}^{i} \frac{V_{k-1} + V_k}{2} \Delta t. \tag{19.7}$$

This summation must be computed for every value of i from 2 to N, thus generating a full vector F_i that represents the anti-derivative of V_i.

time, t (s)	distance, x (m)
0	0
0.19	0.01
0.38	0.04
0.58	0.09
0.77	0.16
0.96	0.25
1.15	0.36
1.34	0.49
1.53	0.65
1.73	0.82
1.92	1.01
2.11	1.22

Exercises 19:

1. Do an internet search for historical data for the last 20 years of the daily Dow Jones Industrial average. (Alternatively, your professor may offer to email it to you.) Import the daily values into Matlab. Write a script to compute the following.

   (a) Write a script to compute the moving average over $m = 40$ days in both directions. (Submit your code and the resultant plot.)

   (b) Write a script to compute the moving standard deviation over $m = 40$ days in both directions. (Submit your code and the resultant plot.)

2. Consider the data for Galileo's inclined plane in the adjacent table. Use Matlab to compute and plot the following.

   (a) Plot the raw data of distance x as a function of time t.

   (b) Use the numeric derivative to compute and plot the velocity as a function of time.

   (c) Use the numeric derivative again to compute and plot the acceleration as a function of time.

time, t (s)	accel., a_x (m/s^2)
0	6.86
0.31	5.77
0.63	2.85
0.94	−0.98
1.25	−4.49
1.56	−6.58
1.87	−6.61
2.19	−4.49
2.50	−0.97
2.81	2.85
3.12	5.78
3.44	6.84

3. An accelerometer is mounted to a 4-wheeled robot. The robot drives forward a distance of approximately 4 m, then backward to its initial position, and the acceleration is measured along the axis of motion. The data is shown in the adjacent table. Use Matlab to compute and plot the following. (Hint: You will need to compute the anti-derivative.)

   (a) Compute and plot the velocity as a function of time. Assume the initial velocity is zero.

   (b) Compute and plot the position as a function of time. Assume the initial position is zero.

20 Fourier Analysis

20.1 Periodic AC Waveforms

So far, we have treated all AC voltages as pure sine waves. In this lecture, we will expand our definition of AC to include any **periodic function** that repeats itself over a length of time T, such that $V(t + T) = V(t)$. Illustrated in Fig. 20.1, we see that any signal, or **waveform**, can be represented as the superposition of a time-varying **AC component** $V_{AC}(t)$ with a **DC offset** $\langle V \rangle$. Mathematically, this is written as

$$V(t) = V_{AC}(t) + \langle V \rangle. \tag{20.1}$$

Importantly, the DC offset $\langle V \rangle$ is just the average value of the signal computed using an integral formula, similar to average AC power dissipation discussed in Lecture 12. The AC component of the signal $V_{AC}(t)$ can be thought of as a summation of sine and cosine functions. This is the mathematical crux of Fourier analysis that we will discuss in this lecture.

Notes: Joseph Fourier was a famous French mathematician who lived during the first industrial revolution.

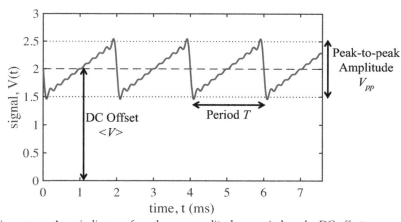

Figure 20.1: A periodic waveform has an amplitude, a period, and a DC offset.

20.2   Fourier Series

In Lecture 13, we used the example of a university marching band to illustrate how a signal, such as your favorite college football fight song, contains various different frequency components. In mathematics, this idea is formalized by **Fourier's theorem**. Fourier's Theorem states that a periodic function $V(t)$ that repeats over a length of time T can be represented as a sum of cosine and sine functions

$$V(t) = V_0 + \sum_{n=1}^{\infty} \left[A_n \cos\left(\frac{2n\pi}{T}t\right) + B_n \sin\left(\frac{2n\pi}{T}t\right) \right], \qquad (20.2)$$

where A_n and B_n are the amplitudes of each frequency component, and A_0 is essentially the DC offset. This summation is known as a **Fourier series**, and the various sine and cosine functions in the sum are often referred to as **frequency components** or **Fourier modes**.

The amplitudes of the modes are related to the periodic function $V(t)$ via the integral formulas

Notes: The integrals of Eqs. (20.3) and (20.4) are analogous to a discrete dot product between $V(t)$ and $\cos\left(\frac{2n\pi}{T}t\right)$ and $\sin\left(\frac{2n\pi}{T}t\right)$, respectively. Thus, it is often said that these formulas are the "projection" of $V(t)$ onto the Fourier modes.

$$A_n = \frac{2}{T} \int_0^T V(t) \cos\left(\frac{2n\pi}{T}t\right) dt, \qquad (20.3)$$

and

$$B_n = \frac{2}{T} \int_0^T V(t) \sin\left(\frac{2n\pi}{T}t\right) dt. \qquad (20.4)$$

Note that the angular frequency of the nth mode is $\omega_n = 2n\pi/T$, and the frequency for the $n = 1$ mode is $\omega_1 = 2\pi/T$, which is known as the **fundamental frequency**. The higher frequencies are all integer multiples of the fundamental frequency, and they are often referred to as **harmonics**.

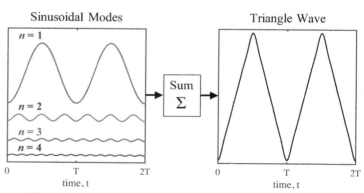

Figure 20.2: Four sine waves with different amplitudes and frequencies are added together to produce a triangle wave.

Illustrated in Fig. 20.2, we see that a triangle wave can created by adding together four cosine waves. The Fourier series formula for the triangle wave is

$$V(t) = \sum_{n=1}^{\infty} \left[\frac{-4}{\pi^2(2n-1)^2} \cos\left(\frac{2\pi(2n-1)}{T}t\right) \right]. \qquad (20.5)$$

For this particular Fourier series, the frequency components are odd integer multiples of the fundamental. However, we still use an integer index n to denote each mode. Also, note that the amplitude of each mode A_n decreases with increasing n. This is necessary for the series to converge.

Example 20.1 demonstrates how to compute and plot a Fourier series in Matlab.

Example 20.1:
Use Matlab to plot the triangle wave given by Eq. (20.5) with only the first four modes. The period should be $T = 2$. Plot two full periods.

Solution: Use linspace() to create a time vector for the independent variable t, and use the zeros() function to initialize a vector containing zeros for the dependent variable $V(t)$. In lines 9 - 11, a for loop will iterate over the mode number n from $n = 1$ to $n = 4$ to compute the sum.

Matlab Script:

```
1 -    clc
2 -    close all
3
4 -    T=2; %period
5 -    t=linspace(0,2*T,10000); %time as independent variable
6 -    Vtri=zeros(1,10000); %initialize V(t) to be all zeros
7
8      %for loop sums up the first 4 modes
9 -    for n=1:4
10 -       Vtri=Vtri-4/pi^2/(2*n-1)^2*cos(2*pi*(2*n-1)*t/T);
11 -    end
12
13 -    figure(2)
14 -    plot(t,Vtri,'lineWidth',2)
15 -    xlabel('time, t')
16 -    ylabel('V(t)')
17 -    set(gca,'FontSize',16,'linewidth',1.5,'FontName',...
18            'Times New Roman')
19 -    set(gcf, 'Position', [100, 100, 450, 450/1.61])
```

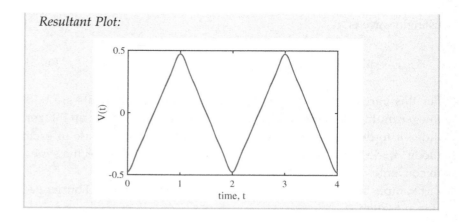

Resultant Plot:

20.3   The Fourier Transform

Any signal can be represented as a single period of a Fourier series, if we simply consider its time domain bounded by a sufficiently large value of T. Making the period T large causes the frequency components $\omega_n = 2n\pi/T$ to become very finely spaced. Taking $T \to \infty$ creates a *continuous* spectrum in the frequency domain. The **Fourier transform** gives us the amplitude and phase of the modes for any analytic function or signal $V(t)$ as a function of a continuous frequency variable ω

$$V(\omega) = \frac{1}{2\pi} \int_{-\infty}^{\infty} V(t) e^{-i\omega t}\, dt. \tag{20.6}$$

Because $e^{-i\omega t} = \cos(\omega t) - i\sin(\omega t)$, the Fourier transform actually contains a real part

$$Re[V(\omega)] = \frac{1}{2\pi} \int_{-\infty}^{\infty} V(t) \cos(\omega t)\, dt, \tag{20.7}$$

and an imaginary part

$$Im[V(\omega)] = -\frac{1}{2\pi} \int_{-\infty}^{\infty} V(t) \sin(\omega t)\, dt. \tag{20.8}$$

These equations are similar to the integral formulas given by Eqs. (20.3) and (20.4) for A_n and B_n and provide the same information. Using the $e^{-i\omega t}$ in Eq. (20.6) simply gives us with a nice compact notation that is more amenable to algebraic manipulation. Furthermore, as we saw in Lecture 12, complex numbers have two degrees of freedom that allows us to information about both amplitude and phase.

The **amplitude** of each frequency component is given by the formula

$$|V(\omega)| = \sqrt{Re[V(\omega)]^2 + Im[V(\omega)]^2}, \qquad (20.9)$$

which has units of amplitude per frequency, so it is often referred to as the **spectral density** of the function. The **phase** of each frequency component is given by the formula

$$\phi(\omega) = \arctan\left(\frac{Im[V(\omega)]}{Re[V(\omega)]}\right). \qquad (20.10)$$

Notes: Recall that ω is the frequency in radians/s and f is the frequency in Hz. In other words, both represent frequency but with different units.

Shown Fig. 20.3, the vibration of a baseball bat is measured and recorded using strain gauges. The Fourier transform gives us the amplitude and phased of the signal as a function of frequency. A large peak appears in the amplitude plot at the resonance frequency of the vibrations.

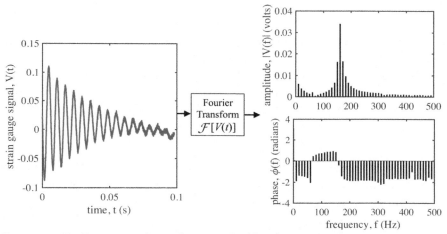

Figure 20.3: The Fourier transform takes a signal $V(t)$ and gives the amplitude and phase of its frequency components. It is often said that the signal is mapped from the time domain into the frequency domain.

20.4   The Discrete Fourier Transform

The Fourier transform given by Eq. (20.6) is an analytic formula that can be applied to a continuous analytic function $V(t)$. However, the signals we record are not continuous, but rather they are discrete digital data sets. For a discrete signal, we must numerically compute the integral using a Riemann sum. This is known as the **Discrete Fourier Transform (DFT)**.

For example, the real part of the Fourier transform becomes

$$\frac{1}{2\pi}\int_0^{T_{max}} V(t)\cos\omega t\,dt \approx \frac{1}{2\pi}\sum_{i=1}^{N} V_i\cos(\omega_j t_i)\Delta t, \qquad (20.11)$$

where T_{max} is the duration of the data collection, N is the number of data points collected, and Δt is the time between successive sample. Angular frequency becomes the new independent variable, and it is represented by a discrete vector ω_j that has N data points and a domain of $[0, 2\pi f_s]$, where f_s is the sampling frequency.

The Riemann sum can be written as a matrix operation

$$
\begin{bmatrix} V(\omega_1) \\ V(\omega_2) \\ \vdots \\ V(\omega_N) \end{bmatrix} = \begin{bmatrix} F_{11} & F_{12} & \cdots & & F_{1N} \\ F_{21} & \ddots & & & \vdots \\ \vdots & & & & \\ F_{N1} & \cdots & F_{N2187} & \cdots & F_{NN} \end{bmatrix} \begin{bmatrix} V(t_1) \\ V(t_2) \\ \vdots \\ V(t_N) \end{bmatrix}, \quad (20.12)
$$

where a vector containing the discrete signal $V_i = V(t_i)$ is multiplied by a kernel matrix

$$
F_{ij} = \cos(\omega_j t_i) \frac{\Delta t}{2\pi}. \quad (20.13)
$$

The **Fast Fourier Transform (FFT)** is a special computer algorithm that performs the matrix operation given by Eqs. (20.12) very efficiently. Example 20.2 demonstrates how to take the FFT of a discrete signal in Matlab. Importantly, the FFT is only valid up to the Nyquist frequency $f_s/2$, and the portion of the FFT for $f > f_s/2$ must be ignored.

Example 20.2:

A middle C note is played on a pipe organ, and the acoustic waves are recorded as a discrete digital signal using a microphone and an oscilloscope with a sampling frequency $f_s = 100$ kHz. The data has saved as a .CSV (comma separated value) file. Write a Matlab script that computes the amplitude as a function of frequency in Hz using the Fast Fourier Transform fft() function.

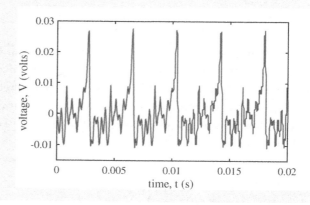

Solution: The Matlab script reads in the discrete signal from the CSV file, and stores it into a vector V. The amplitude of the FFT is simply computed as abs(fft(V)). Importantly, frequency is the new independent variable that will go on the x-axis, and the fft() function does *not* output this data. Rather, we must create a frequency vector using linspace() in line 14 below.

Matlab Script:

```
1 -     clc
2 -     close all
3
4       %read acoustic signal data into matlab
5 -     data = csvread('C_Note.csv')';
6
7 -     t=data(1,:);      %time, s
8 -     VC=data(2,:);     %microphone signal voltage
9
10 -    Tmax=max(t)         %duration, s
11 -    N=length(t)         %number of data pts collected
12 -    fs=1/(t(2)-t(1))    %sampling frequency 1/dt, Hz
13
14 -    f=linspace(0,fs,N); %frequency vector, Hz
15 -    Cfft=abs(fft(VC)/N);%amplitude, Volts
16
17      %Plot the amplitude vs. frequency
18 -    figure(1)
19 -    plot(f,Cfft,'lineWidth',1.5)
20 -    axis([0 2000 0 max(Cfft)*1.1]) %range of 0 to 2000 Hz
21 -    xlabel('frequency, f (Hz)')
22 -    ylabel('amplitude, V(f) (volts)')
23 -    set(gca,'FontSize',16,'linewidth',1.5,'FontName',...
24         'Times New Roman')
25 -    set(gcf, 'Position',  [100, 100, 450, 450/1.61])
```

Resultant Plot:

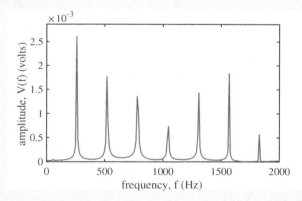

Note that the original waveform was a periodic function with a fundamental frequency of $f_1 = 261$ Hz, as we expect for the middle C note. The spectrum also features spikes at integer multiples of the fundamental frequency, known as **harmonics**.

Exercises 20:

1. Consider the triangle wave given by Eq. (20.5) with a period $T = 2$ ms. Compute the amplitude A_n and frequency ω_n of the $n = 3$ mode.

2. Use Matlab to plot the following Fourier series using only the first 10 modes.

    (a) Triangle Wave: $V(t) = \sum\limits_{n=1}^{10} \left[\frac{-4}{\pi^2(2n-1)^2} \cos\left(\frac{2\pi(2n-1)}{T}t \right) \right]$

    (b) Square wave: $V(t) = \sum\limits_{n=1}^{10} \left[\frac{4}{\pi(2n-1)} \sin\left(\frac{2\pi(2n-1)}{T}t \right) \right]$

    (c) Parabola wave: $V(t) = \sum\limits_{n=1}^{10} \left[\frac{(-1)^n}{n^2} \cos\left(\frac{2\pi n}{T}t \right) \right]$

3. Do an internet search for historical data for the last 20 years of the daily Dow Jones Industrial Average. (Alternatively, your professor may offer to email it to you.) Import the daily values into Matlab. Write a script to compute the amplitude of the FFT as a function of frequency in days^{-1}.

    (a) Plot the $log(|V(f)|)$ as a function of $log(f)$. Only include the portion of the spectrum for $f < f_s/2$.

    (b) Do an internet search for "pink noise". Does the data exhibit pink noise? Explain. (Hint: See Lecture 10 on power laws and log-log plots.)

21 First Order Transient Response

A transient signal $V(t)$ can be recorded from many different physical sources using a wide variety of different sensors. In this lecture and the next, we will discuss several mathematical forms that transient signals often take. Understanding, classifying, and identifying these different **transient responses** allows us to identify underlying physics and predict their behavior under various conditions.

In this lecture, we will look at physical systems that are best described by a **first order!differential equation** of the form

$$\tau \frac{dy}{dt} + y = f(t), \tag{21.1}$$

where τ is a characteristic timescale or **time constant** and $f(t)$ is some time-dependent external stimulus that influences the system. Specifically, we will look at the processes of charging a capacitor and heating a thermocouple.

21.1 Charging of a Capacitor

Recall that a capacitor consists of parallel metal plates that store electrical charge and energy. The charge stored on a capacitor is $q = CV$, where C is the capacitance and V is the voltage across the capacitor. The charge and voltage on a capacitor do appear instantaneously, rather they must be placed there via a transient charging process.

Consider the circuit shown in Fig. 21.1, where the capacitor initially has a charge $q = 0$. At time $t = 0$, the switch is closed, allowing charge to flow through the circuit onto the capacitor at a rate $I = \frac{dq}{dt}$. Applying Kirchhoff's voltage law gives us

$$V_{DC} - R\frac{dq}{dt} - \frac{1}{C}q = 0. \tag{21.2}$$

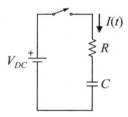

Figure 21.1: The transient charging of the capacitor in this circuit is governed by a first order differential equation.

Equation (21.2) can be rewritten as

$$RC\frac{dq}{dt} + q = CV_{DC}.$$ (21.3)

Comparing this with Eq. (21.1), we see that the characteristic time for this system is $\tau = RC$. You may remember this as the **RC time constant** from physics class. Because the voltage changes abruptly when the switch is closed, the function on the right hand side is technically

$$f(t) = \begin{cases} 0 & \text{if } t < 0 \\ CV_{DC} & \text{if } t \geq 0 \end{cases}.$$ (21.4)

This abrupt change in the external stimulus is known as an **impulse** or **step function**. The solution to Eq. (21.3) is

$$q(t) = CV_{DC}\left(1 - e^{-t/\tau}\right),$$ (21.5)

and the resultant current is

$$I(t) = \frac{dq}{dt} = \frac{V_{DC}}{R}e^{-t/\tau}.$$ (21.6)

Note that as time $t \to \infty$, the charge goes to $q = CV_{DC}$ and the current goes to $I = 0$. Illustrated in Fig. 21.2, we also see that the capacitor is approximately 63% charged at time $t = \tau$ (assuming the switch was closed at time $t = 0$).

Notes: This is the universal definition of a time constant τ for any first order system. It is the amount of time it takes to get 63% of the way from its initial value to steady state.

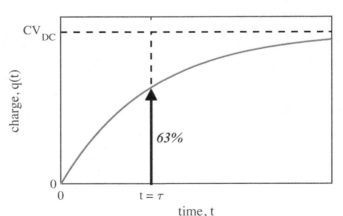

Figure 21.2: The transient response of the RC circuit shown in Fig. 21.1, calculated using Eq. (21.5), is plotted above. The capacitor is 63% charged at time $t = \tau$.

21.2 Response of a Temperature Probe to an Impulse

A temperature probe does not respond instantaneously to a change in temperature, rather it takes time for the probe to heat up or cool down and come into equilibrium with its surroundings. Consider the temperature probe tip in Fig. 21.3. To determine the transient response of such a thermal system, we start by balancing the rate of change of stored thermal energy (or heat) in the probe tip $\dot{Q}_{ST}$ with the net rate that thermal energy flows into the tip $\dot{Q}_{IN} - \dot{Q}_{OUT}$. We will assume that the probe tip has some effective mass m and specific heat c_p, and the net rate of heat flow is given by Newton's law of cooling. This is often referred to as the lumped thermal capacitance model, and it results in the first order differential equation

$$mc_p \frac{dT}{dt} = Ah(T_\infty - T), \qquad (21.7)$$

where A is the effective surface area of the tip, h is the heat transfer coefficient, and T_∞ is the temperature of the surrounding fluid. Equation (21.7) can be rearranged into the form

$$\frac{mc_p}{Ah} \frac{dT}{dt} + T = T_\infty. \qquad (21.8)$$

Comparing this with Eq. (21.1), we see that the **thermal time constant** for this system is $\tau = \frac{mc_p}{Ah}$.

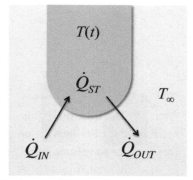

Figure 21.3: The tip of a temperature probe does not respond instantaneously to a change in the surrounding fluid temperature. Rather, it takes time for heat to flow in or out.

Notes: The equation for the thermal time const $\tau = \frac{mc_p}{Ah}$ is not very useful, because m, c_p, A, and h are not well known. To obtain τ, one must measure it by subjecting the probe to an impulse.

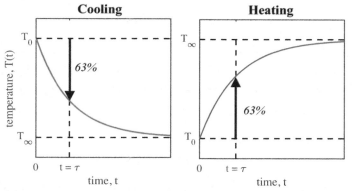

Figure 21.4: The transient response of thermocouple is plotted for both heating and cooling.

Notes: The value of 63% comes from $1 - e^{-1} \approx 0.63$. Thus, it is universally true for any first order system responding to an impulse.

Now, consider the case where the probe is initially at a temperature $T(t = 0) = T_0$, and the temperature of the surrounding fluid is abruptly changed from T_0 to T_∞. (This type of impulse corresponds to quickly submerging a probe that is initially at room temperature into an ice bath or into boiling water.) The solution to Eq. (21.8) is

$$T(t) = T_\infty + (T_0 - T_\infty)e^{-t/\tau}. \qquad (21.9)$$

This equation is valid for both heating and cooling, as shown in Fig. 21.4, and the time constant still corresponds to the amount of time it takes to get to 63% of the way to stead state.

21.3   Response of a Temperature Probe to an Oscillating Temperature

We will now consider the case where the fluid temperature T_∞ is oscillating about some average temperature $\overline{T}$ at some angular frequency ω with an amplitude $|T_\infty|$, such that $T_\infty(t) = \overline{T} + |T_\infty| \sin(\omega t)$. The thermal energy balance equation, Eq. (21.7), then becomes

$$mc_p \frac{dT}{dt} = Ah(\overline{T} + |T_\infty| \sin(\omega t) - T). \tag{21.10}$$

Again, this equation can be rearranged into the form

$$\frac{mc_p}{Ah} \frac{dT}{dt} + T = \overline{T} + |T_\infty| \sin(\omega t). \tag{21.11}$$

Comparing this with Eq. (21.1), we see that the thermal time constant is still $\tau = \frac{mc_p}{Ah}$, and $f(t) = T_\infty(t) = \overline{T} + |T_\infty| \sin(\omega t)$.

We will not go into the detailed solution to Eq. (21.11)—that is better left for a course on differential equations. Instead, we will simply skip ahead to the solution, which is valid for times $t \gg \tau$,

$$T(t) = \overline{T} + \frac{|T_\infty| \sin(\omega t + \phi)}{\sqrt{1 + (\omega \tau)^2}}. \tag{21.12}$$

Inspecting Eq. (21.12), we see that the measured temperature will still oscillate about the same average value $\overline{T}$ at the same frequency ω, as shown in Fig. 21.5. However, the amplitude of the *measured* oscillations will be lower by a factor of

$$\frac{|T|}{|T_\infty|} = \frac{1}{\sqrt{1 + (\omega \tau)^2}}, \tag{21.13}$$

which is also known as the **amplitude ratio**. The measured temperature will also be phase shifted from the actual fluid temperature by an angle

$$\phi = \arctan(-\omega \tau), \tag{21.14}$$

which corresponds to a time lag of

$$\Delta t = \frac{\arctan(-\omega \tau)}{\omega}. \tag{21.15}$$

Notes: The thermal time constant depends heavily on the "surface-to-volume ratio" of the thermal mass. A larger surface-to-volume ratio results in a faster thermal time constant.

Notes: Interestingly, Eqs. (21.13) and (21.14) are similar to the amplitude and phase of an RC low-pass filter. That is because both systems are governed by the same generic first order differential equation: Eq. (21.1).

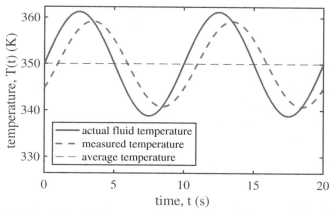

Figure 21.5: The temperature measured with a probe will have a smaller amplitude than the actual temperature of the fluid. It will also have a phase lag.

21.4 Dynamic Calibration and Data Correction

Shown in Fig. 21.5, the measured temperature $T(t)$ is *not* the same as the actual temperature $T_\infty(t)$. This is known as a **systematic error**. A systematic error is when the measured data is not accurate due to some well known physical phenomenon. If one truly understands the reason for a systematic error, then it acceptable to **correct the data** using a mathematical model of the well known physical phenomenon.

In a procedure known as **dynamic calibration**, the time constant τ for the temperature probe (or any other sensor with a first order response) is measured by subjecting the probe to an impulse. Equations (21.13) and (21.14) can then be used to "correct" the measured data. In this case, the formula for the corrected temperature is

$$T_c(t) = \overline{T} + |T_\infty| \sin(\omega t - \phi), \tag{21.16}$$

where $\overline{T}$ is the average of the measured temperature data, ω is the angular frequency of the measured data, and $|T_\infty|$ and ϕ are computed from Eqs. (21.13) and (21.14), respectively. This procedure for correcting data is demonstrated in Example 21.1.

Note that the amplitude and phase both depend on the **dimensionless** or **non-dimensional parameter** $\omega\tau$. For the case of $\omega\tau \ll 1$, the amplitude ratio goes to unity and the phase shift goes to zero. In other words, if the oscillations are slow relative to the thermal response time τ of the probe, then the measured temperature $T(t)$ and actual temperature $T_\infty(t)$ are the same, and the measured data does not need to be corrected.

Notes: Measuring τ by subjecting the system to an impulse is sometimes referred to as **system identification and characterization**.

Notes: For experimentally determining τ, it is usually sufficient to inspect the data and estimate how much time it takes to reach 63%. In Lecture 23, we will discuss a more sophisticated approach.

Notes: If $\tau \ll 1/\omega$, then the measured signal will be accurate, and it is *not* necessary to correct you data.

Example 21.1:

A thermocouple is used to measure the temperature of exhaust from an engine. The thermocouple has a time constant of $\tau = 0.1$ seconds. Plotted below, the measured temperature $T(t)$ oscillates between 541.1 K and 547.3 K at a frequency of 62.8 Hz, which corresponds to an engine speed of 600 RPM.

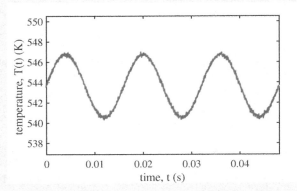

Use Eqs. (21.13) through (21.16) to compute and plot the corrected temperature $T_c(t)$.

Solution: First, we find the average temperature of the measured data to be $\overline{T} = 543.7$ K, which is essentially the midpoint between the maximum and minimum values of the oscillations. In Matlab, this is done by simply taking the mean of the measured temperature data.

Next, the measured peak-to-peak amplitude is $(541.1 - 547.3)$ K $= 6.2$ K. Dividing by 2 gives us the peak amplitude $|T| = 3.1$ K.

Equation (21.13) can then be used to estimate the true amplitude.
$$|T_\infty| = |T|\sqrt{1 + (\omega\tau)^2} = 3.1\text{K}\sqrt{1 + (2\pi(62.8Hz)(0.1s))^2}$$
$$\approx 19.7 \text{ K}$$

The phase can then be calculated from Eq. (21.14).
$$\phi = \arctan(-\omega\tau) = \arctan(-2\pi(62.8Hz)(0.1s)) \approx -88.5°$$

Notes: We must include a factor of 2π to convert from Hz to radians/second.

Lastly, we can use the parameters computed above to plot the corrected temperature vs. time, using Eq. (21.16).

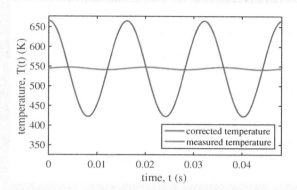

Note that the corrected data has a significantly greater amplitude than the measured data.

Exercises 21:

1. Look at the equation for the thermal time constant of a temperature probe from section 21.2.

   (a) Assume the probe tip is a hemisphere of radius r, and its mass is proportional to its volume, $m = \rho V$. Derive an equation for the time constant in terms of the probe tip radius r.

   (b) Which has a faster response time? A temperature probe with a large tip radius or a small tip radius?

2. A grain of sand is dropped into a lake of water. As it falls, it is subject to the force of gravity $F_g = mg$ and a viscous drag force proportional to velocity $F_d = \gamma v$.

   (a) Use Newton's Second Law, $F = m\frac{dv}{dt}$ to write down a first order differential equation governing the velocity v.

   (b) Rearrange the equation so it is in the same form as Eq. (21.1).

   (c) Derive a formula for the time constant τ in terms of m and γ.

   (d) Derive a formula for the terminal velocity v_T in terms of m, g, and γ. (The terminal velocity occurs when $\frac{dv}{dt} = 0$.)

3. An HVAC specialist is using a thermocouple to measure the air temperature in a ventilation duct. Using a thermocouple with a time constant of 6.5 s in air, he measures a temperature oscillating sinusoidally between 64 and 68°F at a rate of 0.06 Hz.

(a) Estimate the average temperature $\overline{T}$.

(b) What is the amplitude of the measured temperature oscillations?

(c) Use Eq. (21.13) to calculate compute the correct amplitude $|T_\infty|$.

(d) Use Eq. (21.14) to compute the phase shift ϕ.

(e) Use your answers from above to correct the measured signal. That is, make a plot of the measured temperature $T(t)$ and the corrected temperature $T_c(t)$.

22 Second Order Transient Response

Mechanical systems often exhibit natural vibrations or oscillatory motion. If a system naturally vibrates or exhibits periodic oscillations, then it is likely governed by a **second order differential equation** of the form

$$\frac{d^2y}{dt^2} + 2\zeta\omega_n\frac{dy}{dt} + \omega_n^2 y = f(t), \qquad (22.1)$$

where ω_n is the **natural resonance frequency**, ζ is a non-dimensional parameter known as the **damping ratio**, and $f(t)$ is some transient external stimulus that influences the system.

22.1 Damped Spring-Mass Oscillators

Based on our life experiences, we intuitively know that spring-mass systems tend to oscillate. For a more sophisticated understanding of this phenomenon, we turn to classical Newtonian physics. Consider the generic damped spring-mass system illustrated in Fig. 21.1. The mass will be subject to the following forces:

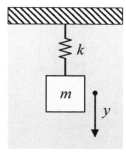

Figure 22.1: A schematic drawing of a spring-mass system.

- **Spring force** - The spring exerts a force that is proportional to the displacement y via Hooke's Law $F_k = -ky$, where k is the spring constant.

- **Viscous damping** - The surrounding fluid (air, water, etc.) also exerts a viscous drag force that is proportional to the velocity $F_\gamma = -\gamma\frac{dy}{dt}$, where γ is the viscous drag force coefficient.

- **External forces** $F(t)$, such as mechanical actuation or acoustic pressure waves, can also effect the system.

Summing up these forces and inserting them into Newton's second law gives us

$$m\frac{d^2y}{dt^2} = -\gamma\frac{dy}{dt} - ky + F(t). \qquad (22.2)$$

Equation (22.2) can be re-arranged into the same form as Eq. (22.1), yielding

$$\frac{d^2y}{dt^2} + \frac{\gamma}{m}\frac{dy}{dt} + \frac{k}{m}y = \frac{F(t)}{m}. \tag{22.3}$$

Comparing Eqs. (22.3) and (22.1), we obtain the following relationships for the damped spring-mass system:

- The **natural resonance frequency:** $\omega_n = \sqrt{\frac{k}{m}}$

- The **damping ratio:** $\zeta = \frac{\gamma}{2}\sqrt{\frac{1}{km}}$

22.2   Response to an Impulse Force

Now, we will consider the case where the external force is abruptly changed, such that

$$F(t) = \begin{cases} 0 & \text{if } t < 0 \\ mg & \text{if } t \geq 0 \end{cases}. \tag{22.4}$$

This is equivalent to holding the mass vertically at a constant position, then abruptly releasing it. Intuitively, we know that such a system will typically bounce up and down, or oscillate. However, the actual behavior is a bit more complex and can change dramatically depending on the relative values of k, m, and γ. The solution to Eq. (22.3) actually takes three different forms depending on the value of the damping ratio ζ.

For $\zeta < 1$, we say the system is **under-damped**, and it exhibits oscillations, as illustrated in Fig. 22.2. The solution takes the form

$$y(t) = Ae^{-\lambda t}\sin(\omega_d t + \phi), \tag{22.5}$$

where A and ϕ are constants that depend on the initial conditions. The system oscillates at the **ringing frequency,** $\omega_d = \omega_n\sqrt{1 - \zeta^2}$, and the **decay constant** , $\lambda = \zeta\omega_n$, determines how quickly the oscillations are damped out.

For $\zeta = 1$, we say the system is **critically damped**, and it does *not* oscillate, as illustrated in Fig. 22.2. One can think of this as the spring-mass system being placed in a container full of honey. The viscous damping force has become so great, that it begins to eliminate the oscillations. The solution takes the form

$$y(t) = Ae^{-\omega_n t} + Bte^{-\omega_n t}, \tag{22.6}$$

where A and B are constants that depend on the initial conditions.

For $\zeta > 1$, we say the system is **over-damped**, and it does *not* oscillate at all, as illustrated in Fig. 22.2. The viscous drag force is so strong that it completely eliminates the oscillations. The solution takes the form

$$y(t) = Ae^{(-\zeta\omega_n + \sqrt{\zeta^2-1})t} + Be^{(-\zeta\omega_n - \sqrt{\zeta^2-1})t}, \tag{22.7}$$

where A and B are constants that depend on the initial conditions. Note that the over-damped case moves much more *slowly* toward its equilibrium at $y = 0$.

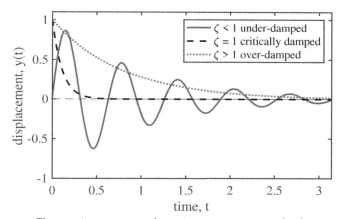

Figure 22.2: The transient response of a spring mass system can be dramatically different depending on the value of the damping ratio ζ.

Notes: We did not include units in Fig. 22.2. This is acceptable when one is merely trying to illustrate the behavior of a mathematical function. However, units must be included when plotting measured data or a predictive theoretical curve.

For many mechanical systems, it is not obvious what values should be used for the various parameters k, m, and γ. In particular, the viscous drag force coefficient γ is very difficult to calculate from first principles. Thus, these parameters must be measured via a process known as **system identification and characterization**. In the following example, we will see how this procedure is applied to a mechanical arm.

Example 22.1:

An engineer wishes to control a mechanical arm using an electric motor, so she develops a mathematical model to help her predict its behavior. The system is illustrated in the free body diagram shown below.

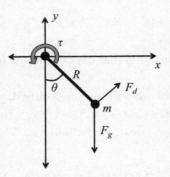

Balancing the angular acceleration with the torque from gravity and the viscous drag force, then applying a small angle approximation $\sin\theta \approx \theta$ yields the equation of motion

$$mR^2\ddot{\theta} = -mgR\theta - \gamma R^2 \dot{\theta},$$

where m is mass, R is the radius of gyration, γ is the viscous drag force coefficient, and g is the acceleration of gravity. She needs to know some of the parameters that go into the mathematical model, so she decides to perform an experiment to characterize the system. The arm is held at a 45 ° angle and released. The resultant oscillations of θ vs. t are recorded and plotted below.

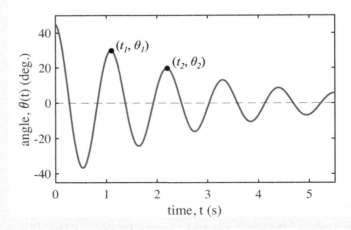

a) The period of the oscillations is determined to be $T = 1.102$s. Calculate the ringing frequency ω_d.

Solution: $\omega_d = 2\pi/T = 5.7$ rad/s

b) The first peak has an amplitude of $\theta_1 = 29.8°$ at time $t_1 = 1.1$ s, and the second peak has an amplitude of $\theta_2 = 19.7°$ at time $t_2 = 2.2$ s. Use the ratio of the peaks to determine the decay constant λ.

Solution: The oscillations are described by Eq. (22.5), and the peaks occur when $\sin(\omega_d t + \phi) = 1$. Thus, Equation (22.5) becomes $\theta_1 = e^{-\lambda t_1}$ and $\theta_2 = e^{-\lambda t_2}$.

Dividing these two equations gives us $\frac{\theta_1}{\theta_2} = e^{-\lambda(t_2 - t_1)}$.

Solving for λ, we obtain $\lambda = \frac{\ln\left(\frac{\theta_1}{\theta_2}\right)}{t_2 - t_1}$.

Substituting in the given values θ_1, θ_2, t_1, and t_2, we obtain $\lambda = 0.375$s^{-1}.

c) The mass of the arm is measured to be $m = 0.2$ kg. Use the formula $\lambda = \gamma/2m$ and the measured decay constant from part b to extrapolate the viscous drag force coefficient γ.

Solution: $\gamma = 2m\lambda = 0.15$kg/s.

d) Use the formula $\omega_d = \sqrt{\frac{g}{R} - \lambda^2}$ and the measured decay constant and ringing frequency to extrapolate the radius of gyration R.

Solution: Solving for R, we obtain $R = \frac{g}{\omega_d^2 + \lambda^2}$.

Substituting in values for g, ω_d, and λ gives us $R = 0.3$ m.

22.3   Response to an Oscillating Force

We will now consider the case where the external force is oscillating at some driving frequency ω, such that $F(t) = F_0 \cos(\omega t)$. Applying Newton's second law give us

$$\frac{d^2y}{dt^2} + \frac{\gamma}{m}\frac{dy}{dt} + \frac{k}{m}y = \frac{F_0}{m}\cos(\omega t). \tag{22.8}$$

The solution to this equation takes the form

$$y(t) = \frac{F_0 \cos(\omega t + \phi)}{m\omega_n^2\sqrt{\left(1 - \frac{\omega^2}{\omega_n^2}\right)^2 + \left(2\zeta\frac{\omega}{\omega_n}\right)^2}}. \tag{22.9}$$

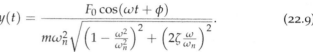

Importantly, the systems oscillates at the same frequency as the driving force ω. The amplitude of the oscillations $|y|$ depends on the driving frequency, and has a maximum value when the driving frequency $\omega = \omega_n\sqrt{1 - 2\zeta^2}$. If the damping ratio is very small, such that $\zeta \ll 1$, then the maximum amplitude occurs at the natural resonance frequency ω_n.

Notes:

- For an impulse force, the system oscillates at its own inherent ringing frequency ω_d.

- For an oscillatory driving force, the system oscillates at the same frequency as the driving force ω.

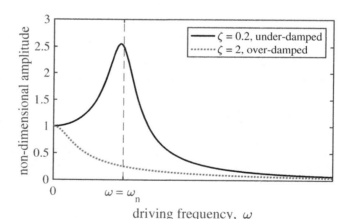

Figure 22.3: The frequency response curve for a damped spring-mass system is plotted for an under-damped case and an over-damped case. The under-damped case has a maximum value for a driving frequency near the natural resonance frequency, ω_n.

The amplitude of the oscillations can be non-dimensionalized as

$$\frac{m\omega_n^2|y|}{F_0} = \frac{1}{\sqrt{\left(1 - \frac{\omega^2}{\omega_n^2}\right)^2 + \left(2\zeta\frac{\omega}{\omega_n}\right)^2}}. \tag{22.10}$$

The non-dimensional amplitude is plotted as a function of the driving frequency ω in Fig. 22.3. This is often referred to as the **frequency response curve** for the system. Note that it peaks near $\omega = \omega_n$ for the under-damped case, and the shape of the curve depends on the value of the damping ratio ζ. If $\zeta > 1$, the frequency response curve does not have a peak, because it is over-damped.

Similar to the band-pass filter in Section 13.5, the under-damped frequency response curve can be characterized by its **full width at half-max (FWHM)** $\Delta\omega$, which is the width of the peak at half of the maximum value. The FWHM is related to the damping ratio and natural resonance frequency via the formula

$$\Delta\omega \approx 2\sqrt{3}\zeta\omega_n. \tag{22.11}$$

Second order systems can also be characterized by measuring their frequency response, then extracting ω_n and $\Delta\omega$ from the data.

The oscillations of the driven system will also be out of phase with the driving force by an amount

$$\phi = \begin{cases} -\arctan\left(\dfrac{2\zeta\frac{\omega}{\omega_n}}{1-\frac{\omega^2}{\omega_n^2}}\right) & \text{if } \omega \leq \omega_n \\[2em] -\pi - \arctan\left(\dfrac{2\zeta\frac{\omega}{\omega_n}}{1-\frac{\omega^2}{\omega_n^2}}\right) & \text{if } \omega > \omega_n. \end{cases} \tag{22.12}$$

The phase is plotted as function of the driving frequency ω in Fig. 22.4.

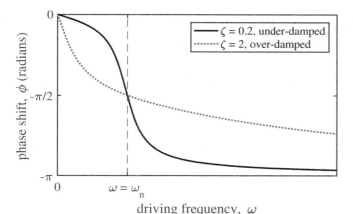

Notes: The phase in Fig. 22.4 always passes through the point $(\omega_n, -\pi/2)$, regardless of the damping ratio ζ. Thus, we say it is a **fixed point**.

Figure 22.4: The periodic oscillations of damped spring-mass system will be out of phase from the driving force by an angle ϕ that depends on the driving frequency ω, as shown above.

Exercises 22:

1. Similar to Example 22.1, a mechanical arm is to be controlled using an electric motor, and the same experiment was performed to
characterize the system.

   (a) The period of the oscillations is determined to be $T = 0.785$s.
   Calculate the ringing frequency ω_d.

   (b) The first peak has an amplitude of $\theta_1 = 18.8°$ at time $t_1 = 0.78$
   s, and the second peak has an amplitude of $\theta_2 = 7.8°$ at time
   $t_2 = 1.57$ s. Use the ratio of the peaks to determine the decay
   constant λ.

   (c) The mass of the arm is measured to be $m = 90$ g. Use the
   formula $\lambda = \gamma/2m$ and the measured decay constant from part b
   to extrapolate the viscous drag force coefficient γ.

   (d) Use the formula $\omega_d = \sqrt{\frac{g}{R} - \lambda^2}$ to extrapolate the radius of
   gyration R.

2. Use calculus to show that Eq. (22.10) has a peak at $\omega = \omega_n\sqrt{1 - 2\zeta^2}$.
(Hint: Take the derivative with respect to ω and set it equal to zero.)

3. A piezoelectric crystal expands and contracts when a voltage is applied to its surface. Applying an AC voltage to the crystal causes
it to vibrate, and the system can be modeled as a driven damped
spring-mass oscillator. The frequency response data is shown in the
adjacent table, with the driving frequency f in kHz and the amplitude of the mechanical oscillations in arbitrary units (a.u.).

   (a) Plot the data in Matlab. Use the '-o' option to make the data
   points be connected by lines.

   (b) Estimate the natural resonance frequency of the system in units
   of kHz. (Assume $\zeta \ll 1$.)

   (c) Estimate the full width at half of the max (FWHM) $\Delta\omega$ in units
   of kHz.

   (d) Use your answers from above to estimate the damping ratio ζ.
   (Recall that ζ is non-dimensional, so make sure all of the units
   cancel out.)

freq. f (kHz)	amp. $|y|$ (a.u.)
39.86	45
39.93	52
39.99	61
40.05	73
40.12	92
40.18	118
40.24	151
40.31	166
40.37	145
40.43	112
40.49	88
40.56	70
40.62	58
40.68	50
40.75	43
40.81	38
40.87	34
40.94	31

23 Data Processing (Part II)

Shown below, the Eagle Nebula (M16) is an enormous cloud of gas in outer space, spanning a distance of over 10^{16} m. In the adjacent image, a typical atmospheric cloud seen on here on earth has a similar structure and appearance, despite having a much smaller size of approximately 10^2 m. Both cloud structures are similar, because they both have a similar **Reynolds Number**.

Figure 23.1: The Eagle Nebula in outer space (left) is much greater in size than a typical atmospheric cloud on earth (right). However, the gas density of the nebula is only $\rho \sim 10^9$ m^{-3}, while the density of the atmospheric cloud is $\rho \sim 10^{25}$ m^{-3}. Thus, both have similar Reynolds numbers. (Images courtesy of NASA.)

In this lecture, we will see how various non-dimensional parameters can be used to characterize and compare systems with similar physics, in spite of vastly different magnitudes in their characteristic parameters. We will also see how non-linear data sets can be transformed to become linear for the sake of parameter extrapolation.

23.1   Non-dimensional Parameters

In aerospace engineering, it is common to perform wind tunnel tests on experimental aircraft. However, a full size aircraft is almost never placed in a wind tunnel. Building such a large wind tunnel and operating it at fast airspeeds would cost millions, or even billions of dollars. Thus, aerodynamic tests are typically performed on smaller scale models, and non-dimensional parameters such as the Reynolds number are employed. Listed below are several non-dimensional parameters commonly used in **aerodynamic** theory and testing.

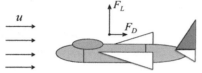

Figure 23.2: An aircraft is subject to a lift and drag force.

- The **Reynolds Number** $Re = \frac{\rho u L}{\mu}$, where ρ is the density of the fluid, u is the airspeed or "free stream velocity", L is some relevant length scale for the system, and μ is the dynamic viscosity of the fluid. Theoretically, the Reynolds number quantifies the ratio of inertial forces to viscous forces. For the sake of experimentation, it can be used to make a scale model with a smaller value of L. One must simply ensure the Reynolds number of the scale model system matches the Reynolds number of the full sized system. This is typically achieved by using water or compressed air to increase the density ρ.

Notes: At room temperature, the dynamic viscosity of air is $\mu_{air} \approx 17 \times 10^{-6}$ Ns/m^2, and the dynamic viscosity of water is $\mu_w \approx 900 \times 10^{-6}$ Ns/m^2.

- The **Lift Force Coefficient** $C_L = \frac{F_L}{\frac{1}{2}\rho u^2 A}$, where F_L is the lift force on an airplane wing, ρ is the density of the fluid, u is the airspeed or "free stream velocity", and A is the surface area of the airplane wing. The lift force coefficient quantifies the lift force on an aircraft wing, as illustrated in Fig. 23.2. In a wind tunnel experiment, it is typically measured as a function of the Reynolds number.

- The **Drag Force Coefficient** $C_D = \frac{F_D}{\frac{1}{2}\rho u^2 A}$, where F_D is the drag force on an airplane wing, ρ is the density of the fluid, u is the airspeed or "free stream velocity", and A is the surface area of the airplane wing. The drag force coefficient quantifies the drag force on an aircraft wing, as illustrated in Fig. 23.2. In a wind tunnel experiment, it is typically measured along side the lift force as a function of the Reynolds number.

Notes: The dynamic viscosity μ and the speed of sound c both change dramatically with temperature.

- The **Mach Number** $Ma = \frac{u}{c}$, where u is the airspeed or "free stream velocity" and c is the speed of sound. The Mach number is used to quantify speed in supersonic flows, or aircraft travelling faster than the speed of sound.

Non-dimensional parameters are also widely used in heat transfer theory and experiments. Listed below are several non-dimensional parameters commonly used in **heat transfer**.

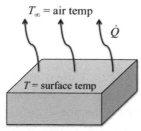

T_∞ = air temp

$\dot{Q}$

T = surface temp

Figure 23.3: Heat is transferred from a solid material into a surrounding fluid.

- The **Biot Number** $Bi = \frac{hL}{k_S}$, where h is the convective heat transfer coefficient, L is some relevant length scale for the system, and k_S is thermal conductivity of the *solid* material. The Biot Number is used to determine the mode of heat transfer for a solid material dissipating heat into a surrounding fluid, as illustrated in Fig. 23.3. For a small Biot Number $Bi \ll 1$, heat flows easily through the solid, and cooling is limited by how quickly heat can be convected away by the fluid. It is often used to determine when a theoretical "lumped thermal capacitance" model, such as the one in Section 21.2, is valid.

- The **Nusselt Number** $Nu = \frac{hL}{k_F}$, where h is the convective heat transfer coefficient, L is some relevant length scale for the system, and k_F is thermal conductivity of the *fluid*. The Nusselt Number is used quantify the heat transfer coefficient h. Experimentally, it is a dependent variable analogous to the lift and drag force coefficients, and it is typically measured as a function of the Reynolds number.

- The **Prandtl Number** $Pr = \frac{C_p \mu}{k_F}$, where μ is the dynamic viscosity of the fluid, C_p is the specific heat of the fluid, and k_F is thermal conductivity of the fluid. Unlike the other parameters listed above, the Prandtl number is an inherent property of any fluid, as it is calculated solely from fluid properties without any physical length scale or external velocity imposed upon it.

In Lecture 13 on RC filters and Lecture 21 on first order response, we saw the non-dimensional parameter $\omega\tau$ that characterizes the response of a system driven by some external stimulus oscillating at a driving frequency ω. In the following example, we will see how data from a low-pass filter can be **collapsed onto a single curve** using non-dimensional parameters.

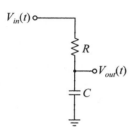

$V_{in}(t)$

R

$V_{out}(t)$

C

Figure 23.4: A low-pass filter is constructed using a resistor and capacitor.

Example 23.1:

An RC low-pass filter circuit, shown in the adjacent figure, was constructed and tested using $C = 10$nF and an input amplitude $|V_{in}| = 10$V. The frequency response curve was measured using $R = 2000\Omega$ and again using $R = 4700\Omega$. Shown below, the measured amplitude data is plotted on top of the theoretical formulas for both resistors. (See Lecture 13 on analog filters for more details.)

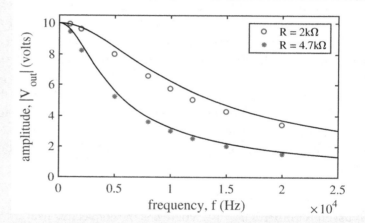

Make a plot of the data using non-dimensional parameters. Specifically, plot the amplitude ratio $|V_{out}|/|V_{in}|$ as a function of the non-dimensional parameter ωRC along with the theoretical curve.

Solution: We define new variables $y = |V_{out}|/|V_{in}|$ and $x = \omega RC$. Thus, the theoretical curve given by Eq. (13.1) becomes $y = \frac{1}{\sqrt{1+x^2}}$.

The measured data is also transformed, and new variables "x2000", "x4700", etc. are defined in the Matlab script on lines $22 - 26$.

Matlab Script:

```
4      C=10e-9;      %capacitance, Farads
5      Vin=10;       %amplitude of Vin, volts
6      f=[20 15 12 10 8 5 2 1]*1000; %driving freq., Hz
7      w=2*pi*f;         %convert Hz to rad/s
8
9      %R1 = 2kOhm
10     R1=2000;
11     V1=[3.4 4.28 5.04 5.74 6.56 8 9.6 9.92];      %measured amplitude, Volts
12
13     %R2 = 4.7kOhms
14     R2=4700;
15     V2=[1.51 2.02 2.52 3 3.6 5.24 8.24 9.44]; %measured amplitude, Volts
16
17     %Non-dimensional theoretical curve
18     x=linspace(0,7,1000);
19     y=1./sqrt(1+x.^2);
20
21     %Non-dimensional measured data
22     x2000=w*2000*C;
23     y2000=V1/Vin;
24
25     x4700=w*4700*C;
26     y4700=V2/Vin;
27
28
29     figure(1)
30     plot(x2000,y2000,'o',x4700,y4700,'*',x,y,'k','linewidth',1.5)
31     xlabel('\omegaRC')
32     ylabel('|V_{out}|/|V_{in}|')
33     legend('R = 2k\Omega','R = 4.7k\Omega')
```

Resultant Plot:

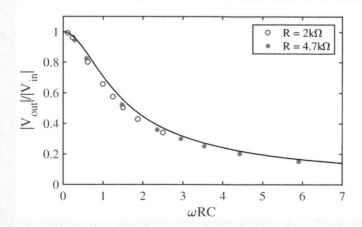

Note that the theoretical curve is the same for both data sets. Thus, we say the data has been "collapsed onto a single curve".

23.2   Linearizing Data

In Lecture 10, we saw how a log-log plot will linearize non-linear data if it is governed by a power law. We will now generalize this beyond power laws to any non-linear function. The overarching idea is to define new variables x' and y', in terms of the original variables x and y, such that the new variables obey a linear function $y' = mx' + b$. The new variables will be a function of the old variables, where $x' = f(x,y)$ and $y' = g(x,y)$. This is often referred to as **transforming** or **mapping** the original data into a new space.

Consider for example, the first order response of a thermal system subject to an impulse that was presented in Lecture 21. The temperature vs. time data theoretically obeys the exponential decay function

$$T(t) = T_\infty + (T_0 - T_\infty)e^{-t/\tau}. \tag{23.1}$$

where T_∞ is the temperature of the surrounding fluid, T_0 is the initial temperature, and τ is the thermal time constant. Rearranging (23.1) and taking the natural log of both sides gives us

$$\ln\left(\frac{T(t) - T_\infty}{T_0 - T_\infty}\right) = \frac{-t}{\tau}. \tag{23.2}$$

Notes: Not all thermal systems will obey this type of exponential decay. Equation (23.1) is only valid if the Biot number is small, $Bi \ll 1$.

We can then define a new variable

$$y(t) = \ln\left(\frac{T(t) - T_\infty}{T_0 - T_\infty}\right). \tag{23.3}$$

Using this transformation, Eq. (23.2) becomes a linear function of time

$$y = \left(-\frac{1}{\tau}\right)t, \tag{23.4}$$

where the slope of the line will be $m = -1/\tau$.

As we will see in the following example, measured temperature vs. time data can be linearized by this transformation, if it obeys the scaling predicted by Eq. (23.1). Furthermore, applying a linear curve fit will allow us to extrapolate the time constant τ.

Example 23.2:

A small metal work piece was initially at $T_0 = 9.6°C$, then placed in hot water at $T_\infty = 35.7°C$. The temperature vs. time data was recorded using an A/D and plotted below.

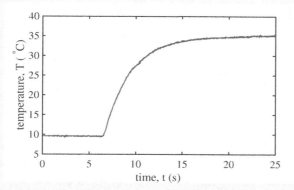

The data appears to obey the theoretical first order response predicted by Eq. (23.1). In Matlab, linearize the data using Eq. (23.3) to confirm the theory is correct. Then, apply a linear curve fit to extrapolate the time constant τ.

Matlab Script:

```
1 -     clc
2 -     close all
3
4 -     data=csvread('thermo_data.csv'); %import data
5 -     t=data(:,1);     %time data
6 -     Temp=data(:,2); %temperature data
7
8 -     T0=9.6;      %initial temperature
9 -     Tinf=35.7;    %ambient fluid temperature
10
11 -    y=log((Temp-Tinf)/(T0-Tinf)); %transform the data to linearize it
12
13      %apply a linear fit to the linear portion of the data
14 -    c=polyfit(t(240:540),y(240:540),1)
15
16      %generate vectors containing the linear fit
17 -    tfit=linspace(0,25,1000);
18 -    yfit=c(1)*tfit+c(2);
19
20      %extrapolate the time constant from the slope
21 -    slope=c(1)
22 -    tau=-1/slope    %time constant, s
23
24 -    figure(1)
25 -    plot(t,y,'lineWidth',2);
26 -    hold on
27 -    plot(tfit,yfit,'linewidth',1);
28 -    hold off
29 -    xlabel('time, t (s)')
30 -    ylabel('y(t)')
31 -    set(gca,'FontSize',16,'linewidth',1.5,'FontName',...
32          'Times New Roman')
33 -    set(gcf, 'Position', [100, 100, 450, 450/1.61])
```

Resultant Plot:

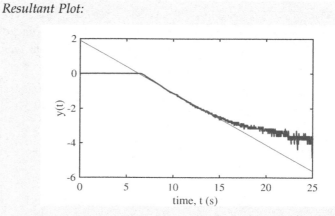

Importantly, only a portion of the transformed data is linear. This is because the theory is only valid for times after the sample was placed in the hot water. The transformed data also becomes scattered and non-linear at longer times, because the difference $T(t) - T_\infty$ is less than the digital resolution of the A/D and the electronic noise in the measurement.

Notes: Slope is defined as "rise-over-run" $\Delta y/\Delta x$, so it will have units of the "rise" Δy divided by units of the "run" Δx.

The linear portion of the transformed data was determined by inspecting the data, and a linear curve fit was applied. The slope from the curve fit is $m = -0.301$ s^{-1}. Thus, the time constant can be extrapolated to be $\tau = 3.32$ s.

Exercises 23:

1. An engineer wishes to measure the aerodynamic drag on an automobile. The full size car has a characteristic length of $L = 2$ m and a cross sectional area of $A = L^2 = 4$ m^2. It travels at a typical highway speed of $u = 30$ m/s.

    (a) Calculate the Reynolds number for the full size car traveling through atmospheric air.

    (b) There is no wind tunnel large enough to fit the full size car, so the engineer will perform a water tunnel test to measure the drag force on a smaller scale model of the car. The water tunnel moves at $u' = 10$m/s. What should the length scale L' of the scale model car be in order to match the Reynolds number of the full size car?

    (c) In the water tunnel, the drag force on the scale model is measured using a load cell to be $F_D' = 254$ N. Calculate the measured drag force coefficient. (Assume $A' = (L')^2$.)

    (d) Use the measured drag force coefficient obtained from the water tunnel measurement to determine the drag force on the full size car traveling through air at a typical highway speed of $u = 30\text{m/s}$.

2. The number of bacteria cells in a petri dish grows exponentially in time, such that

$$N(t) = N_0 e^{rt},$$

where N_0 is the initial population at $t = 0$, and r is the growth rate.

(a) Take the natural log of both sides of the equation. Show that the transformation $y(t) = \ln(N(t))$ yields a linear relationship between y and t.

(b) Consider the theoretical relationship between y and t that you just derived. What is the theoretical slope and intercept in terms of N_0 and r?

(c) The population N of a bacteria cell culture in a petri dish is measured as a function of time t, and the resultant data is contained in the adjacent table. Use Matlab to make a plot of the population as a function of time.

(d) Linearize the data using the transformation you just derived. Use Matlab to make a plot of the linearized data and apply a linear curve fit.

(e) Use the slope of the linear curve fit to extrapolate the growth rate r. (Be sure to include the correct units.)

t (hrs)	N (million)
0	0.098
8	0.200
16	0.408
24	0.832
32	1.698
40	3.467
48	7.077

3. The kinematic viscosity of engine oil v decreases with temperature T. Theoretically, it obeys an Arrhenhius relationship of the form

$$v = v_0 \exp\left(\frac{E_A}{k_B T}\right),$$

where v_0 is the exponential pre-factor, E_A is the activation energy, and $k_B = 1.38 \times 10^{-23}$ J/K is the Boltzmann constant.

(a) Use algebra to show that the transformation $y' = \ln(v)$ and $x' = 1/T$ will linearize measured data. (This type of transformation is known as an **Arrhenius plot**.)

(b) What is the theoretical slope and intercept of the linearized data in terms of v_0, E_A, and k_B?

24 Uncertainty

Like any man-made artifact, all sensors and transducers have subtle flaws, and no measurement is ever 100% correct. Furthermore, it is very rare to repeat an experiment and obtain the exact same result every time.

We will begin this lecture with two important definitions.

- **Experimental Error** is the difference between the true value and the measured value $\epsilon = x_{true} - x_{measured}$.

- **Experimental Uncertainty** is an *estimate* of the experimental error.

We typically do not know the error, because we do not know the true value x_{true}. We can only estimate it.

24.1 Instrumental Uncertainty

All sensors and transducers have subtle flaws that lead to experimental error. **Instrument uncertainty** U_I is an estimate of the experimental error associated with a particular sensor. The instrument uncertainty is estimated by looking at the manufacturer's **data sheet**, which contains many different parameters and technical specifications for the sensor. One typically reads through the data sheet and looks for the sources of uncertainty, which often have a "$\pm$" symbol in front of them. These different sources of uncertainty are then added in quadrature using the formula

Notes: When using Eq. (24.1), it is critical that all the parameters being added in quadrature have the same units. i.e., One does not simply add a percentage to a voltage.

$$U_I = \sqrt{U_{RES}^2 + U_L^2 + U_Z^2 + U_H^2 + \cdots}. \tag{24.1}$$

The parameters U_{RES}, U_L, U_Z, U_H, etc. represent the various sources of error for the sensor. The data sheet may contain only one or two of these, or it may contain all of them plus many more, depending on how thorough the manufacturer wishes to be.

Listed below are some common sources of error for a sensor or transducer.

- **Resolution Uncertainty** U_{RES} is typically half of the smallest division. You may have learned this in chemistry class when measuring volume with a graduated cylinder, such as the one illustrated in Fig. 24.1. It also applies to digital electronics, such as a digital multimeter (DMM), where voltages are represented by discrete integer values.

- **Linearity** or **Non-linearity** U_L is when a linear calibration is applied, but the true relationship is not quite linear.

Figure 24.1: A graduated cylinder, used for measuring volume, has tick marks indicating 2 mL divisions, so it has resolution uncertainty $U_{RES} = 1$ mL. Thus, we would say the measured volume shown above is $V = 25 \pm 1$ mL.

- **Zero-shift** or **Offset** U_Z is when the intercept of the calibration equation tends to drift randomly in time. A common example of this is a scale for measuring mass. With nothing on the scale, it should read zero, but this is not always the case.

- **Hysteresis** U_H is when the sensor outputs a different value for the same measured parameter depending on whether the parameter is being increased or decreased, as illustrated in Fig. 24.2.

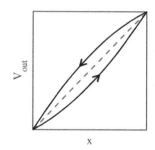

- **Accuracy** U_A is often used as a generic catch-all for the various sources of uncertainty. It is usually determined by using the sensor to measure a known standard, such as the temperature of ice water and directly calculating the experimental error.

Figure 24.2: An illustration of hysteresis for an analog sensor.

If you are lucky, the sensor manufacturer will specify a **total error band (TEB)** in their data sheet. This means that they have already accounted for all of the sources of error, substituted them into Eq. (24.1), and calculated the instrument uncertainty for you. The total error band is often given as a percentage of the sensor's total **range** or **full scale span (FSS)**.

Importantly, the instrument uncertainty U_I for a sensor can be calculated before it is even purchased, and this is often a critical step in experimental design. In the following example, we will see how to estimate the instrument uncertainty for an analog pressure sensor.

Notes: **Full scale span (FSS)** is synonymous with sensor **range**.

Example 24.1:

An engineer is planning to purchase a TE Connectivity 4515-DS5B002DP analog pressure sensor from digikey.com to be used in a wind tunnel experiment. Before spending money on the sensor, the engineer downloads the data sheet to check the manufacturers specifications.

To follow along with this example, you will need to go to digikey.com (a popular electronics vendor), search for the part number, and download the datasheet.

This pressure sensor is from the "4515" family of pressure sensors from TE Connectivity. On page 15 of the datasheet, there is a guide that explains the other numbers and letters in the part number.

- **DS**5B002DP – The "DS" means this model has "Dual Sideports", where the pressure tubes from the Pitot probe will be connected.

- DS**5**B002DP – The "5" means that the sensor needs 5V DC to power it.

- DS5B**002**DP – The "002" means that the sensor has a pressure range of 2 inches of water.

- DS5B002**D**P – The "D" means that the sensor will measure the pressure *difference* ΔP between the two dual sideports.

- DS5B002D**P** – The "P" means that the sensor has "thru hole" electrical connections, which means it can be plugged into a standard breadboard or connected to wires with female DuPont pins.

We can estimate the instrument uncertainty U_I from information in the datasheet, as well. On page 4, we see that the total error band (TEB) for sensor ranges of 4 inches of water and below is $\pm 2\%$ of the span. The span—or range—for this model is 2 inches of water. Thus, the instrument uncertainty is

$$U_I = (0.02)(2 \text{ in. } H_2O) = 0.04 \text{ in. } H_2O.$$

24.2   Repeatability Uncertainty

It is very rare to repeat an experiment and obtain the exact same result every time. Rather, measurements are randomly scattered about some mean $\bar{x}$ as illustrated in Fig. 24.3. The standard deviation s is then used to quantify the width of the scatter.

The **repeatability uncertainty** U_R is used to quantify how repeatable the measurement is, and it is calculated using the formula

$$U_R = \frac{(t_{\nu,\%C})s}{\sqrt{N-1}}, \tag{24.2}$$

where $t_{\nu,\%C}$ is the Student's t-factor, N is the number of times the experiment was performed, and $\nu = N - 1$ is the number of degrees of freedom.

The **Student's t-factor** $t_{\nu,\%C}$ can be found in **Appendix A** at the end of the book. It is used to adjust for the percent confidence %C. For example, choosing 90% means that 90% of experimental measurements will be between $\bar{x} - U_R$ and $\bar{x} + U_R$. This is often referred to as a **confidence interval**, and it is graphically represented by the **error bar** at the bottom of Fig. 24.3. The following example illustrates how to compute the repeatability uncertainty.

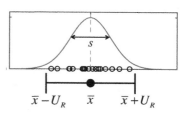

Figure 24.3: When an experiment is repeated N times, the measured from each experiment are typically scattered about some mean value $\bar{x}$.

Notes: In the very first lecture, we said that the scientific method relies on **repeatable observations**. If a phenomenon is only observed $N = 1$ times, the repeatability uncertainty U_R is infinite. This is often referred to as **anecdotal evidence**, and we should be infinitely skeptical of its validity.

Example 24.2:

An engineer is measuring the pressure using the TE Connectivity 4515-DS5B002DP sensor from the previous example. She performs the experiment $N = 8$ times, resulting in the data shown in the adjacent table. Compute the mean $\bar{p}$, standard deviation s, and repeatability uncertainty U_R for the experimental results.

Solution: The mean and standard deviation can be computed in Matlab using the mean() and std() function, yielding $\bar{p} = 1.3338$ in.H_2O and $s = 0.09812$ in.H_2O.

```
p=[1.38 1.35 1.19 1.33 1.22 1.45 1.29 1.46]
pbar=mean(p)
s=std(p)
```

Referring to the Student's t-table in **Appendix A**, we go down to the row corresponding to $\nu = N - 1 = 7$, then over to the column for 90% confidence, which gives us $t_{7,90\%} = 1.895$. We then substitute these values into Eq. (24.2).

$$U_R = \frac{(1.895)(0.098 \text{ in.}H_2O)}{\sqrt{8-1}} = 0.070 \text{ in.}H_2O.$$

pressure, p (in.H_2O)
1.38
1.35
1.19
1.33
1.22
1.45
1.29
1.46

Notes: The standard deviation and the repeatability uncertainty should have the same units as the data. Don't forget to include the units!

24.3   Total Uncertainty

After repeating an experiment N times, the mean of the data $\bar{x}$ is computed along with the standard deviation s and repeatability uncertainty U_R, as we saw in the previous example. However, the repeatability of the experiment is not the only source of uncertainty. The sensor itself also has an inherent instrument uncertainty U_I. The final step is then to compute the **total uncertainty** by adding the instrument and repeatability uncertainties in quadrature

$$U_{TOT} = \sqrt{U_I^2 + U_R^2}. \tag{24.3}$$

The final result of the experimentation is then reported as $\bar{x} \pm U_{TOT}$.

Notes: Important values determined from an experiment should be proudly reported in the abstract or summary section at the beginning of a research publication or technical memo.

As a convention, the total uncertainty is typically rounded to just two significant digits, and the mean is rounded to the least significant digit of the uncertainty. Consider the following.

- $x = 4.01382 \pm 0.241934$ is **incorrect**. Remember, uncertainty is only an *estimate* of the error, and an estimate is not accurate enough to include any decimals beyond 0.24.

- $x = 4.01382 \pm 0.24$ is still **incorrect**. It does not make sense to include any more digits after 4.01. The true value of x could be as low as 3.77382 or as large as 4.25382. If we are that uncertain about the digits in the ones and tenths place, why bother including digits that are 1000× smaller? The digits 0.00382 are **not significant** relative to the uncertainty 0.24. That is, $0.00382 << 0.24$.

- $x = 4.01 \pm 0.24$ is the **correct** way to report the value.

Example 24.3:
Compute the total uncertainty for the series of wind tunnel experiments using the TE Connectivity 4515-DS5B002DP analog sensor in the previous examples.

Solution: Use the results from the previous two examples and substitute into Eq. (24.3) to compute the total uncertainty.
$$U_{TOT} = \sqrt{(0.04 \text{ in.} H_2O)^2 + (0.07 \text{ in.} H_2O)^2} = 0.081 \text{ in. } H_2O$$

Thus, we report the average result of the $N = 8$ wind tunnel experiments as $\boxed{p = 1.334 \pm 0.081 \text{ in. } H_2O}$

Exercises 24:

1. Look up the data sheet for the analog pressure sensor from Example 24.1. Consider the following different versions of the sensor, which all have different ranges.

   (a) Estimate the instrument uncertainty for the 4515-DS5B010DP.

   (b) Estimate the instrument uncertainty for the 4515-DS5B020DP.

   (c) Estimate the instrument uncertainty for the 4515-DS5B030DP.

   (d) An engineer wishes to measure a pressure of approximately 16 in. H_2O. Which sensor would be best for this application?

2. An Omron Electronics D6T-44L-06H MEMS thermal infrared temperature sensor is used to measure the surface temperature of a mechanical part in an experiment. Go to digikey.com, search the part number, download the data sheet, and answer the following questions. (Alternatively, your professor may choose to send you the data sheet.)

temperature, T (°C)
47.3
50.4
49.0
47.9
48.3

   (a) What type of digital communication does the D6T-44L-06H sensor use to output its data?

   (b) What is the sensor's range for "object temperature detection"?

   (c) What is the sensor's "object temperature output accuracy" U_A?

   (d) What is the sensor's "Temperature resolution" U_{RES}?

   (e) Use the previous two answers to compute the instrument uncertainty U_I.

   (f) The sensor is used to measure the surface temperature, and the experiment is repeated $N = 5$ times. Use the data in the adjacent table to compute the mean, standard deviation, and repeatability uncertainty at 90% confidence.

   (g) Compute the total uncertainty U_{TOT} for the experimental results. Report the final result as $\overline{T} \pm U_{TOT}$ with the correct significant digits.

distance, x (m)
8.85
9.49
9.46
9.47
9.17
8.42
9.19
9.55
9.09
9.31

3. A Terabee Tower Evo 60m LIDAR module (part number TR-TW-EVO60M-4) is used in a self-driving autonomous vehicle to detect and measure the distance to nearby objects and obstacles. Go to mouser.com, search the part number, download the data sheet, and answer the following questions. (Alternatively, your professor may choose to send you the data sheet.)

(a) What is the range for the sensor?

(b) What is the resolution U_{RES} for objects within 14 m distance?

(c) What is the accuracy U_A for objects within 14 m distance?

(d) Use the previous two answers to compute the instrument uncertainty U_I.

(e) The sensor is used to perform an experiment $N = 10$ times. Use the data in the adjacent table to compute the mean, standard deviation, and repeatability uncertainty at 95% confidence.

(f) Compute the total uncertainty U_{TOT} for the experimental results. Report the final result as $\overline{T} \pm U_{TOT}$ with the correct significant digits.

25 Error Propagation

Measured parameters are often used in theoretical calculations. For example, consider a measured parameter $x = \bar{x} \pm U_x$ that is substituted into a function $f(x)$. The uncertainty in x will result in an uncertainty in $f(x)$, as the experimental error propagates through the calculation, as illustrated in Fig. 25.1.

In this lecture, we will see how to estimate the uncertainty in a function $f(x, y, z)$ that results from the uncertainty in the input parameters x, y, and z.

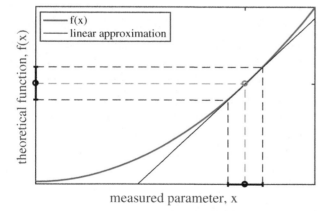

Notes: Error bars can be included in a plot by using the errorbar() function in Matlab.

Figure 25.1: A parameter on the x-axis has some uncertainty, such that its true value lies somewhere in the rage $\bar{x} \pm U_x$. Plugging this parameter into a function $f(x)$ results in a new with a value somewhere in the range $f \pm U_f$. This range of possible values is represented by the **error bars** on the input x and output $f(x)$.

25.1   Linear Approximations of Uncertainty

Consider a function $f(x,y,z)$ that has multiple input parameters x, y, and z. Each of the input parameters has an uncertainty associated with it such that

$$x = \bar{x} \pm U_x$$
$$y = \bar{y} \pm U_y \tag{25.1}$$
$$z = \bar{z} \pm U_z.$$

The uncertainty in each parameter can be projected onto the function using a **linear approximation**. For example, the uncertainty is x gets projected onto the function via $\Delta f = f'(\bar{x}) \cdot U_x$. This is graphically illustrated in Fig. 25.1.

The total uncertainty that results from the uncertainty in all three input parameters is obtained by adding the linear approximations for all three in quadrature

$$U_f = \sqrt{\left(U_x \frac{\partial f}{\partial x}\Big|_{\bar{x},\bar{y},\bar{z}}\right)^2 + \left(U_y \frac{\partial f}{\partial y}\Big|_{\bar{x},\bar{y},\bar{z}}\right)^2 + \left(U_z \frac{\partial f}{\partial z}\Big|_{\bar{x},\bar{y},\bar{z}}\right)^2}. \tag{25.2}$$

Importantly, one must take the **partial derivatives** of the function, then substitute the mean values $\bar{x}$, $\bar{y}$, and $\bar{z}$. The following example demonstrates how to properly use Eq. (25.2).

Notes: When using Eq. (25.2), always check that the parameters added in quadrature have the same units.

Example 25.1:
An engineer is using an analog pressure sensor with a Pitot probe to measure the airspeed of a small, unmanned aircraft. The pressure sensor measures a differential pressure of $\Delta p = 611 \pm 75$ kPa. Fluctuations in the atmospheric temperature and pressure result in an uncertainty in the air density, such that $\rho = 1.22 \pm 0.18$.

Compute the airspeed u and its uncertainty U_u. Express your answer in the form $u = \bar{u} \pm U_u$ with the correct number of significant digits and appropriate units.

Solution: First, we use the simplified Bernoulli equation, given by Eq. (2.2), to determine the airspeed.

$$u = \sqrt{\frac{2\Delta p}{\rho}} = \sqrt{\frac{2(611\text{ Pa})}{(1.22\text{ kg/m}^3)}} = 31.911 \text{ m/s}.$$

Next, we use Eq. (25.2), to estimate the uncertainty in the calculated airspeed.

$$U_u = \sqrt{\left(\frac{\partial u}{\partial(\Delta p)}U_{\Delta p}\right)^2 + \left(\frac{\partial u}{\partial \rho}U_\rho\right)^2}.$$

Taking the partial derivatives and substituting them into the formula gives us

$$U_u = \sqrt{\left[\sqrt{\tfrac{2}{\rho}}\left(\tfrac{1}{2\sqrt{\Delta p}}\right)U_{\Delta p}\right]^2 + \left[\tfrac{-\sqrt{2\Delta p}}{2\rho^{3/2}}U_\rho\right]^2}.$$

This formula is a bit unwieldy, so we will simplify it algebraically before typing it into the calculator.

$$U_u = \sqrt{\tfrac{U_p^2}{2\rho\Delta p} + \tfrac{\Delta p U_\rho^2}{2\rho^3}}.$$

Ahhh, that's much better! Substituting in values gives us $U_u = 3.093$ m/s. Thus, we report our final answer as

$$\boxed{u = 31.9 \pm 3.1 \text{ m/s}}$$

25.2 Special Rules for Addition, Subtraction, Multiplication, and Division

For simple addition, subtraction, multiplication, and division of parameters, Equation (25.2) reduces to two simple formulas.

- For **addition and subtraction** $f(x,y,z) = x \pm y \pm z$, we simply add the absolute uncertainties in quadrature. This works for any combination of addition or subtraction applied to any number of terms.

$$U_f = \sqrt{U_x^2 + U_y^2 + U_z^2} \tag{25.3}$$

- For **multiplication and division** $f(x,y,z) = xyz$, $f(x,y,z) = \tfrac{xy}{z}$, etc., we simply add the **relative uncertainties** in quadrature. This works for any combination of multiplication or division applied to any number of terms.

$$\frac{U_f}{f(\bar{x},\bar{y},\bar{z})} = \sqrt{\left(\frac{U_x}{\bar{x}}\right)^2 + \left(\frac{U_y}{\bar{y}}\right)^2 + \left(\frac{U_z}{\bar{z}}\right)^2} \tag{25.4}$$

Note that we must multiply both sides by $f(\bar{x},\bar{y},\bar{z})$ to get the absolute uncertainty U_f.

$$U_f = f(\bar{x},\bar{y},\bar{z})\sqrt{\left(\frac{U_x}{\bar{x}}\right)^2 + \left(\frac{U_y}{\bar{y}}\right)^2 + \left(\frac{U_z}{\bar{z}}\right)^2} \tag{25.5}$$

Example 25.2:

A mechanical assembly of cylindrical parts is illustrated below with dimensions and tolerances.

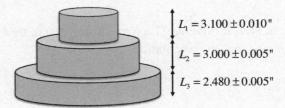

$L_1 = 3.100 \pm 0.010''$

$L_2 = 3.000 \pm 0.005''$

$L_3 = 2.480 \pm 0.005''$

Solution: First, we add up the individual lengths to compute the total height of the stack $h = 3.100'' + 3.000'' + 2.480'' = 8.580''$.

Next, we simply add the uncertainties (tolerances) of the dimensions in quadrature.

$$U_f = \sqrt{(0.01'')^2 + (0.005'')^2 + (0.005'')^2} \approx 0.012''$$

Thus, we report the total expected height for the assembly as

$$\boxed{h = 8.580 \pm 0.012''}.$$

Example 25.3:

A rectangular steel workpiece is illustrated below with dimensions and tolerances.

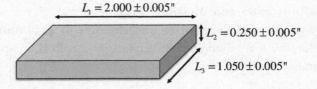

$L_1 = 2.000 \pm 0.005''$

$L_2 = 0.250 \pm 0.005''$

$L_3 = 1.050 \pm 0.005''$

The density of the steel is $\rho_s = 0.28 \pm 0.02$ lbs./in^3. Calculate the mass of the workpiece and its uncertainty. Express you answer in the form $m = \overline{m} \pm U_m$ with the correct number of significant digits and appropriate units.

Solution: First, we multiply the density times the volume to compute the mass $m = \rho_s L_1 L_2 L_3 \approx 0.147$ lbs.

Next, we add the relative uncertainties of the dimensions and density in quadrature and multiply by the calculated total mass of the sample.

$$U_m = m\sqrt{\left(\frac{U_{\rho_s}}{\rho_s}\right)^2 + \left(\frac{U_{L_1}}{L_1}\right)^2 + \left(\frac{U_{L_2}}{L_2}\right)^2 + \left(\frac{U_{L_3}}{L_3}\right)^2} \approx 0.011 \text{ lbs.}$$

Thus, we report the total expected mass for the workpiece as

$$\boxed{m = 0.147 \pm 0.011 \text{ lbs.}}$$

Exercises 25:

1. Algebraically show that Eq. (25.2) reduces to . . .

    (a) Eq. (25.3) for $f(x,y,z) = x - y + z$.

    (b) Eq. (25.4) for $f(x,y,z) = xyz$.

    (c) Eq. (25.4) for $f(x,y,z) = \frac{xy}{z}$.

2. A culture of bacteria are growing in a Petri dish. Similar to Exercise 23.2, the number of cells grows exponentially, such that

$$N(t) = N_0 e^{rt},$$

    where N_0 is the initial population at $t = 0$, and r is the growth rate.

    (a) Derive a formula for the uncertainty in the population U_N in terms of t, N_0, r, U_{N_0}, and U_r.

    (b) The initial population is $N_0 = 9800 \pm 1200$, and it grows at a rate of 0.078 ± 0.013 per hour. Calculate the population at $t = 40$ hours along with its uncertainty. Express your answer in the form $N = \overline{N} \pm U_N$ with the correct number of significant digits.

    (c) Which contributes more to the uncertainty in the population? The uncertainty in the initial population N_0, or the uncertainty in the growth rate r?

3. Calculate the following parameters along with their uncertainties. Express your answers in the form $x \pm U_x$. Do not forget to include units!

(a) An LED light uses $i = 20 \pm 3.4$ mA of current at a voltage of $V = 4.8 \pm 0.2$ V. Calculate the power $\dot{q}_{LED}$ and its uncertainty.

(b) Two locomotive trains are on a head-on collision course. One is traveling at a speed of $v_1 = 85 \pm 9$ mph and the other at $v_2 = 65 \pm 6$ mph. Calculate the speed at which they collide $v_{TOT} = v_1 + v_2$ and its uncertainty.

(c) A chemist measures the mass of a solution to be $m = 10.32 \pm 0.04$ g and the volume to be $V = 9.4 \pm 0.5$ mL. Calculate the density of the solution ρ and its uncertainty.

26 Introduction to Probability

In the previous lecture, we introduced the idea of **confidence intervals**. That is, the repeatability uncertainty had a percent confidence associated with it, which represented the **probability** of measuring a value between $\bar{x} - U_R$ and $\bar{x} + U_R$. In the remaining lectures, we will see how probability and statistics can be used to interpret data and make informed engineering decisions.

26.1 Formal Logic

Probability and statistics deals with an ensemble of N observations, and it considers what percentage of the observations will yield a certain outcome. As human beings, we naturally live in the moment and only consider a single event that is occurring in front of us. Therefore, it is necessary for us to expand our minds and consider N events occurring simultaneously. A number of useful tools involving **formal logic** have been developed to help us with this unnatural mode of thought.

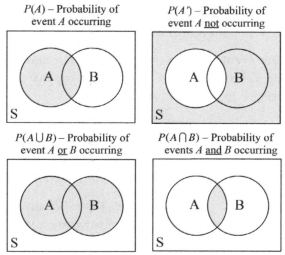

Figure 26.1: Venn diagrams provide a visual representation of abstract ideas involving probability.

Shown in Fig. 26.1, Venn diagrams are our first tool for understanding probability theory. The visual cortex is the largest part of the human brain, and visual graphics can help us understand abstract mathematical ideas. The shaded areas represent the probability of specific events or combinations of events occurring, and the surrounding rectangle includes and exhaustive set of all possible outcomes S.

Listed below are some important definitions and mathematical formulas corresponding to the Venn diagrams in Fig. 26.1.

Notes: Here we see much of the same formal logic that was introduced in Lecture 16: "NOT", "AND", and "OR".

- $P(A)$ is the **probability** of event A occurring, and it always takes values between 0 and 1 or 0 and 100%.

- $P(A') = 1 - A$ is the probability of event A *not* occurring. It is sometimes referred to as the **compliment** of A.

- The **union** $P(A \cup B)$ is the probability of either event A **or** event B occurring. The formula to calculate the union is

$$P(A \cup B) = P(A) + P(B) - P(A \cap B), \qquad (26.1)$$

where $P(A \cap B)$ is the intersection of the events (see next bullet point).

- The **intersection** $P(A \cap B)$ is the probability of events A **and** B both occurring. The formula to calculate the intersection is

$$P(A \cap B) = P(A)P(B|A), \qquad (26.2)$$

Notes: The conditional probability is incredibly important to science, engineering, and medicine, because it helps establish a causal link between two events. i.e., the probability of getting lung cancer given that you worked with asbestos.

where $P(B|A)$ is the conditional probability of A given B (see next bullet point).

- The **conditional probability** $P(B|A)$ is the probability of events B occurring given that event A has occurred.

- **Independent events** are two events that do not have any influence on each other, such that $P(B|A) = P(B)$. We say that the probability of B occurring is independent of A.

- **Mutually exclusive** means that events A and B cannot possibly occur simultaneously, such that $P(A \cap B) = 0$. For example, a blizzard and a 90° F heat wave are mutually exclusive events. There is no possible overlap, as illustrated by the Venn diagram in Fig. 26.2.

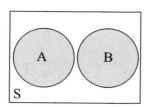

Figure 26.2: Events A and B are mutually exclusive.

- A set of events $S = \{A_1, A_2, A_3, \ldots A_N\}$ is **exhaustive** if it includes all possible outcomes. If the set is exhaustive, then $\sum_{n=1}^{N} P(A_n) = 1$. Importantly, the set $\{A, A'\}$ is exhaustive, because A either occurs or it does not occur. There are no other possibilities. As a corollary, we can infer $P(A) + P(A') = 1$.

The following two examples demonstrate how to apply these ideas and formulas.

Example 26.1:

Consider a standard deck of 52 playing cards. The deck contains 26 red cards, 4 Kings. Two of the Kings are red, and two are black. For the following calculations, assume the deck has been shuffled to a completely random order.

a) What is the probability of drawing two kings in a row from the deck?

Solution: We essentially want the probability that the first card is a King and the second card is a King. The word "and" means that this is an intersection $P(A \cap B) = P(A)P(B|A)$.

The probability of drawing the first King is $P(A) = 4/52$, and the probability of drawing a second King *given the first was King* is $P(B|A) = 3/51$. (After the first King has been drawn, there are only 3 Kings out 51 remaining cards.)

Substituting into the equation for the intersection gives us $P(A \cap B) = \left(\frac{4}{52}\right)\left(\frac{3}{51}\right) = \frac{1}{221} \approx 0.45\%$.

b) What is the probability of drawing a red card or a King?

Solution: The word "or" means that this is a union of events $P(R \cup K) = P(R) + P(K) - P(R \cap K)$.

The probability of a red card is $P(R) = 26/52 = 1/2$, and the probability of a King is $P(K) = 4/52$.

We must also consider the fact that 2 out of 52 cards in the deck are red Kings (both red and Kings), or $P(R \cap K) = 2/52 = 1/26$. This intersection must be subtracted, lest we double-count the red Kings.

$$P(R \cup K) = P(R) + P(K) - P(R \cap K) = \frac{1}{2} + \frac{1}{13} - \frac{1}{26} = \frac{7}{13}.$$

Example 26.2:
An engineer is designing a photovoltaic array of solar panels. Three solar cells must be connected in series to achieve the desired voltage. His first design—Design (a)—is shown below.

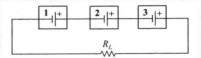

a) The probability that an individual solar cell lasts more than 5 years is 90%. If any individual solar cell fails, then the entire array will fail, because they are connected in series. Calculate the probability that the array will last more than 5 years. (Assume that the failure of individual solar panels are independent events.)

Solution: For the array to survive, all three solar panels must survive. That is, panel 1 and 2 and 3 must all survive more than 5 years.

$$P(A_1 \cap A_2 \cap A_3) = P(A_1)P(A_2)P(A_3) = (0.9)^3 = \boxed{0.729 \text{ or } 72.9\%}$$

b) The engineer decides that probability of design (a) surviving is unacceptably low, so he decides to add a redundant row of three more solar panels connected in parallel to the first. His second design—Design (b)—is shown below. Calculate the probability that Design (b) lasts more than 5 years.

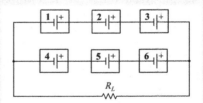

Solution: For Design (b) to survive, the top row must survive *or* the bottom row must survive. From the previous design, we saw that the probability of the top row was
$$P(TOP) = P(A_1)P(A_2)P(A_3) = (0.9)^3 = 0.729$$

By symmetry, we can assume that the probability of the bottom row surviving is the same. Thus the probability of the top row or the bottom row surviving is given by the union.

$$P(TOP \cup BOT) = P(TOP) + P(BOT) - P(TOP \cap BOT)$$

$$= 0.729 + 0.729 - (0.729)^2 \approx \boxed{0.927 \text{ or } 92.7\%}$$

c) In a stroke of genius, the engineer realizes that Design (b) can be further improved by adding just a few more inexpensive wires. The third design, Design (c) shown, below has both redundant rows and columns. Calculate the probability that Design (c) lasts more than 5 years.

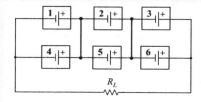

Solution: Design (c) will fail if both of the cells in a column fail. For example, probability that both the cells in the first column fail is $P(A_1' \cap A_4') = (1 - 0.9)^2$.

The probability that both the cells in the first column do *not* fail is $P[(A_1' \cap A_4')'] = 1 - (1 - 0.9)^2$.

Extending this logic to the entire array, we want the probability that the first column does not fail *and* the second column does not fail *and* the third column does not fail. Using formal logic, this is written as $P[(A_1' \cap A_4')' \cap (A_2' \cap A_5')' \cap (A_3' \cap A_6')']$.

Thus, the probability that Design (c) lasts more than 5 years is given by the mathematical formula

$$P[(A_1' \cap A_4')'] \cdot P[(A_2' \cap A_5')'] \cdot P[(A_3' \cap A_6')']$$

$$= [1 - (1 - 0.9)^2]^3 \approx \boxed{0.970 \text{ or } 97\%}.$$

Notes: Design (c) is 2× *less* likely to fail than Design (b). Subtle differences like the two extra wires can make significant improvements for little extra cost. Ideas such as this are what separate average engineers from extraordinary engineers.

26.2   DeMorgan's Law

For the simple union of two events, we add the probabilities and subtract the intersection to avoid double counting it. This is illustrated graphically in Fig. 26.3, where the area of the sideways snowman shape represents the union of the two events. Simply adding the area of the two individual circles will result in double-counting the intersection.

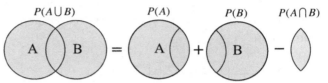

Figure 26.3: Equation (26.1) is represented graphically. The intersection must be subtracted, otherwise it will be counted twice.

The union of three or more events becomes considerably more difficult to calculate. That is because there are multiple intersections that must be subtracted, as illustrated in Fig. 26.4. For calculating the union of three or more events, it is much easier use **DeMorgan's Law**

$$P(A \cup B \cup C) = 1 - P(A' \cap B' \cap C'). \tag{26.3}$$

This formula essentially replaces the unions with intersections, and it can be extended to more than three events.

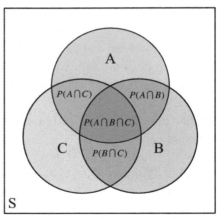

Figure 26.4: Three separate events have 4 possible intersections. The number of possible intersections will increases substantially with the number of events.

The following example illustrates the application of DeMorgan's Law to an everyday concern.

Example 26.3:

A farmer is concerned that the corn is not getting enough water, and he is considering turning on his irrigation system. There is a 60% chance of rain on Monday, a 50% chance of rain on Tuesday, and a 30% chance of rain on Wednesday. Calculate the probability that it will rain on either Monday, Tuesday, or Wednesday.

Solution: First, we recognize the word "or", which implies the union of the three events $P(M \cup T \cup W)$. As we saw in the Venn diagram in Fig. 26.4, there are 4 different intersections that must be subtracted from the sum. To avoid this cumbersome calculation, we will simply apply DeMorgan's Law.
$$P(M \cup T \cup W) = 1 - P(M' \cap T' \cap W')$$

Assuming they are independent events, this becomes
$$P(M \cup T \cup W) = 1 - P(M')P(T')P(W')$$

Substituting in values gives us
$$P(M \cup T \cup W) = 1 - (1 - 0.6)(1 - 0.5)(1 - 0.3) = 0.86$$

Thus, we say that there is an $\boxed{86\%}$ chance that it will rain on either Monday, Tuesday, or Wednesday.

Notes: Without using DeMorgan's Law, we would have to subtract off four different intersections:

- Rain on Mon. and Tues.
- Rain on Mon. and Weds.
- Rain on Tues. and Weds.
- Rain on all 3 days

Exercises 26:

1. A university has 256 students studying mechanical engineering. Among these students, 106 are female, 42 are grad students, and 18 are female grad students. Calculate how many of the students are . . .

   (a) undergraduates.
   (b) female undergraduates.
   (c) male undergraduates.
   (d) either female or grad students.

2. A bag containing 30 resistors comes off the assembly line at a factory, and 9 of them are defective. An inspector draws 10 at random from the bag. Find the probability that none of the inspected resistors are defective if ...

   (a) each one is replaced and the bag is shaken before a new one is drawn.
   (b) they are not replaced before a new one is drawn.

3. A coffee maker will fail if the heater burns out, the water pump fails, or the drip nozzle becomes clogged. There is a 10% probability the heater will burn out, a 15% chance the water pump will fail, and a 30% chance the drip nozzle will become clogged within the next 4 years.

   (a) Calculate the probability that the coffee maker will fail within the next 4 years.

   (b) Calculate the probability that the coffee maker does not fail within the next 4 years.

4. It is the early spring of 218 BC, and Hannibal has amassed an army of over 100,000 men and 38 war elephants in Southern France. They are preparing to invade Italy and conquer Rome. To reach the Italian peninsula, they can take mountain pass A, or mountain pass B, or mountain passes C and D, which are connected in series via a small mountain village.

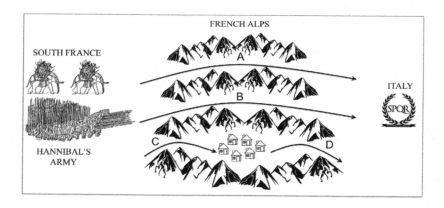

Mountain passes A and B each have a 50% chance of being impassable due to a blizzard. Mountain passes C and D each have a 60% chance of being impassable due to damaged roads. What is the probability that Hannibal and his army will be able to pass through the Alps into Italy? (Assume they will take any route that is passable.)

27 Discrete Probability Distributions

Think about the physics of rolling dice or a spinning roulette wheel. Like any mechanical system, the outcome of these actions is governed Newton's equations of motion with appropriate initial conditions. However, the equations of motion are so complex that we would need to write a computer program to solve them. Furthermore, the solution calculated by such a computer program is incredibly sensitive to the initial conditions—change the initial angular momentum of the die or the roulette wheel by 0.0001%, and you will get a completely different answer. This profound sensitivity to initial conditions for a complex system is often referred to as **chaos**. For such systems, it is necessary to use the mathematics of **probability and statistics** to understand and predict the behavior.

This lecture will cover **discrete probability distributions**, where $P(n)$ is the probability of observing some discrete value n. For example, a single six-sided die has a discrete probability distribution, as illustrated in Fig. 27.1.

n	1	2	3	4	5	6
$P(n)$	1/6	1/6	1/6	1/6	1/6	1/6

Figure 27.1: The discrete probability distribution for a fair six-sided die is given above. It is called a "fair" die distribution, because there is an equally probability of observing (rolling) any of the 6 possible values.

27.1 Normalization, Expected Value, and Standard Deviation

If a discrete probability distribution contains all possible outcomes, then it is an exhaustive set, and the sum of all the probabilities must be unity. We say that the distribution is **normalized** if

$$\sum_n P(n) = 1. \tag{27.1}$$

For the fair die distribution shown in Fig. 27.1, the sum of all possible values is one, and the distribution is normalized.

All probability distributions have a **mean** or **expected value** given by the formula

$$\langle n \rangle = \sum_n nP(n). \qquad (27.2)$$

The expected value can be denoted as $\langle n \rangle$ or $\bar{n}$. Importantly, the expected value is not always an integer, nor does it necessarily correspond to measurable value. For example, the expected value for the fair dice distribution is

$$\langle n \rangle = \sum_{n=1}^{6} nP(n) = 1(1/6) + 2(1/6) + \ldots + 6(1/6) = 3.5.$$

Clearly, it is impossible to roll a 3.5 with a single roll. However, if we take the average of *many* rolls, we expect it to be $\langle n \rangle \approx 3.5$.

Discrete probability distributions also have a **standard deviation** s given by the formula

$$s = \sqrt{\sum_n (n - \langle n \rangle)^2 P(n)}. \qquad (27.3)$$

The standard deviation is typically denoted as s or σ. The **variance** of the distribution is simply the standard deviation squared s^2. For our fair die distribution, we have

$$s = \sqrt{\sum_{n=1}^{6} (n - \langle n \rangle)^2 P(n)}$$
$$= \sqrt{(1-3.5)^2\frac{1}{6} + (2-3.5)^2\frac{1}{6} + \ldots (6-3.5)^2\frac{1}{6}}$$
$$\approx 1.708.$$

Note that one must compute the expected value or mean $\langle n \rangle$ first before computing the standard deviation. Similar to the mean, if we rolled the dice many times and took the standard deviation of the data, we expect it to be $s \approx 1.71$.

27.2   The Binomial Distribution

Often times, an experimental result is binary in nature. That is, the outcome is either true or false—the event A either occurs or does not occur. For example, a mechanical part either passes or fails a stress test, or the roll of a dice either comes up as a 6 or it does not. Consider performing such an experiment N times.

The **binomial distribution** gives us the probability of observing the event A a total of v_A times out of N trials, given that the probability of the event occurring in an individual trial is $P(A)$. The formula for the binomial distribution is

$$P[v_A, N, P(A)] = \left[\frac{N!}{(N - v_A)! v_A!}\right] P(A)^{v_A} [1 - P(A)]^{N - v_A}. \quad (27.4)$$

The first term represents the number combinations that an event can occur v_A times out of N trials. The second term $P(A)^{v_A}$ is the intersection of the event occurring v_A times. The third term $[1 - P(A)]^{N - v_A}$ is the intersection of the event *not* occurring $N - v_A$ times.

The following example illustrates how the binomial distribution is used.

Example 27.1:
What is the probability that exactly 2 out of 5 dice rolls will be a 4? (Assume it is a fair dice.)

Solution: The event A that we are considering is the dice roll coming up as a $n = 4$. The phrase "2 out of 5" tells us that $v_A = 2$ and $N = 5$. For a fair die, the probability of rolling a 4 is $P(A) = 1/6$.

$$P[2, 5, 1/6] = \left[\frac{5!}{(5-2)! 2!}\right] (1/6)^2 [1 - 1/6]^{(5-2)} = 10(1/36)(5/6)^3$$

≈ 0.161 or 16.1%.

Alternatively, we can use the binopdf() function in Matlab.

```
>> binopdf(2,5,1/6)

ans =

    1.6075e-01
```

Importantly, the binomial distribution is properly normalized, such that

$$\sum_{\nu_A=0}^{N} P[\nu_A, N, P(A)] = 1. \qquad (27.5)$$

Furthermore, the different values of ν_A are mutually exclusive. For example, it is impossible for 2 out of 5 dice rolls to come up as 4 and 5 out of 5 to come up as 4. These two facts will be important in the following example.

Example 27.2:

Dr. Shaquille O'Neal was a talented professional basketball player, who had difficulty shooting free throws. His career free throw average was only 52.7%, which means the probability of him making an individual free throw was $P(A) = 0.527$.

a) Calculate the probability that Shaq will make less than 3 out of 11 free throws.

Solution: The phrase "less than 3 out of 11" tells us that $N = 11$ and $\nu_A < 3$. That is, ν_A can take values 0 or 1 or 2. This is a union, so we must add the three probabilities given by the binomial distribution together. (We do not have to subtract off any intersections, because the different values of ν_A are always mutually exclusive.)

$$P(\nu_A < 3) = P[0, 11, 0.527] + P[1, 11, 0.527] + P[2, 11, 0.527]$$

This can be computed using the binopdf() function in Matlab.

```
>> binopdf(0,11,0.527)+binopdf(1,11,0.527)+binopdf(2,11,0.527)

ans =

   2.1617e-02
```

Alternatively, we could use the binocdf() function in Matlab, which computes the sum of the probabilities from $\nu_A = 0$ to a specified value.

```
>> binocdf(2,11,0.527)

ans =

   2.1617e-02
```

Both approaches give us $P(\nu_A < 3) \approx \boxed{0.0216 \text{ or } 2.16\%}$.

b) Calculate the probability that Shaq will make at least 3 out of 11 free throws.

Solution: The phrase "at least 3 out of 11" tells us that $N = 11$ and $v_A \geq 3$. That is, v_A can take values 3 or 4 or 5 or 6 or 7 or 8 or 9 or 10 or 11. Again, we must add the probabilities given by the binomial distribution together.

However, there is a shortcut that will save us time. Equation (27.5) tells us that $\sum_{v_A=0}^{11} P[v_A, N, P(A)] = 1$. Therefore, it follows that

$$P(v_A \geq 3) = P[3, 11, 0.527] + P[4, 11, 0.527] + \ldots + P[8, 11, 0.527]$$
$$= 1 - P(v_A < 3).$$

Using our answer from part a), we have

$$P(v_A \geq 3) = 1 - P(v_A < 3) \approx \boxed{0.978 \text{ or } 97.8\%}.$$

27.3 Poisson Statistics

To measure the probability that an event A occurs, we simply perform N trials and count the number of times v_A that the event is observed. The **measured probability** is simply calculated using

$$P(A) = \frac{v_A}{N}. \tag{27.6}$$

However, there is an inherent statistical uncertainty associated with this measurement technique. Consider measuring the probability that the roll of a fair die will come up as four. If we only perform the experiment $N = 6$ times, there is a significant probability that none of the rolls would be a 4, and $v_A = 0$, which would give us a measured probability $P(A) = 0$.

This "probability of measuring different probabilities" is essentially given by the binomial formula, which is plotted in Fig. 27.2 with v_A on the horizontal axis for different values of N.

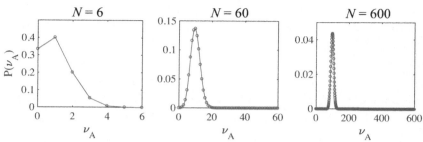

Figure 27.2: The binomial probability distribution is plotted as a function of ν_A for different number of trials N. As N increases, a sharp peak emerges at $\nu_A = N \cdot P(A)$.

Examine Fig. 27.2 and consider the following.

- For a small number of trials with $N = 6$, there is a good chance we could measure something dramatically different from $P(A) = 1/6$.

- If more trials are performed, such that $N = 60$, a clear peak emerges near $\nu_A = 10$, indicating it is likely we will measure something close to $P(A) = 10/60 = 1/6$.

- If many more trials are performed, such that $N = 600$, an even sharper peak emerges near $\nu_A = 100$, indicating it is even more likely we will measure something close to $P(A) = 100/600 = 1/6$.

The width of the peaks in Fig. 27.2 quantifies the statistical uncertainty in the measured probability. According to **Poisson statistics**, the width decreases as $1/\sqrt{N}$, and the measured probability has a statistical uncertainty or **Poisson error** of

$$U_{P(A)} = \sqrt{\frac{P(A)}{N}}. \tag{27.7}$$

Importantly, the uncertainty in the measured probability *decreases* as the number of trials N is increased. This is known as **statistical convergence**.

The following example demonstrates how Poisson statistics can be applied to mechanical testing.

Example 27.3:

A mechanical engineer performs a stress test on a batch of manufactured parts, and 15 out of 75 of the parts fail.

a) Calculate the probability and uncertainty in the probability of the part failing. (Express your answer in the form $P(F) = P \pm U_P$ with the correct number of significant digits.)

Solution: The probability is calculated using Eq. (27.6).

$$P(F) = \frac{\nu_F}{N} = \frac{15}{75} = 0.2 \text{ or } 20\%$$

The uncertainty in the probability is calculated using Eq. (27.7).

$$U_P(F) = \sqrt{\frac{P(F)}{N}} = \sqrt{\frac{0.2}{75}} \approx 0.0516 \text{ or } 5.16\%$$

The measured probability of the part failing the stress test is

$$\boxed{P(F) = 20.0 \pm 5.2\%} \text{ or } \boxed{P(F) = 0.200 \pm 0.052}.$$

b) The engineer decides that the uncertainty U_P is too large. Estimate how many parts should be tested to lower the uncertainty by a factor of 10.

Solution: We want the new uncertainty to be $10\times$ lower, such that $U'_{P(F)} = U_{P(F)}/10$.

This is achieved by increasing the number of trials from N to N'. Using Eq. (27.7) gives us

$$\sqrt{\frac{P(F)}{N'}} = \frac{1}{10}\sqrt{\frac{P(F)}{N}}$$

For the sake of a simple estimate, we will assume that the measured probability $P(F)$ will not change by an appreciable amount. Thus, it will cancel from both sides of the equation.

$$\sqrt{\frac{1}{N'}} \approx \frac{1}{10}\sqrt{\frac{1}{N}}$$

Solving for N', we have $N' \approx 10^2 N$, so we would need to increase the number of trials from $N = 100$ to $N' = 10,000$.

Notes: As more trials are performed and N is increased, we get diminishing returns on the decrease in U_P. Eventually, it becomes cost prohibitive to try to decrease U_P any further.

27.4   Histograms

Our previous examples have all involved a discrete probability distribution $P(n)$, where n was an integer. However, the observable n does not necessarily have to be an integer. Consider measuring the electrical resistance of manufactured resistors. Like any manufactured part, the resistors will not all be identical. Shown in Fig. 27.3 is a **histogram** of the measured resistance for $N = 400$ different resistors.

Notes: Choosing the correct number of bins is important when making a histogram plot. Typically, you vary the number of bins until the bin with the greatest number of counts contains 50 to 100 counts.

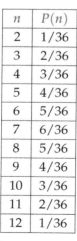

Figure 27.3: Electrical resistances were measured for a set of $N = 400$ resistors. The distribution of measured values is plotted using the histogram() function in Matlab with 20 bins.

The **histogram** breaks up the range of measured values into small intervals known bins, and plots the number of measured values within each bin as a bar graph. This type of plot can be made in Matlab using the histogram() function.

Exercises 27:

n	$P(n)$
2	1/36
3	2/36
4	3/36
5	4/36
6	5/36
7	6/36
8	5/36
9	4/36
10	3/36
11	2/36
12	1/36

1. For a *pair* of six-sided dice, it is possible to roll any number from 2 to 12. Please use the fair dice probability distribution shown in the adjacent table to answer the following.

(a) Show that the distribution is properly normalized.

(b) Compute the theoretical mean, or expected value.

(c) Compute the theoretical standard deviation.

(d) Plot the distribution using the bar() function in Matlab.

(e) Obtain a pair of dice. Roll them $N = 100$ times, and enter the value of each roll into a spreadsheet. Make a histogram of the data. Compute the mean and standard deviation of the measured data, and compare it to the theoretical values.

2. Use the binomial probability distribution function to answer the following.

  (a) For a fair six-sided die, what is the probability that exactly 4 out of 12 rolls will be a 2?

  (b) For a fair coin, what is the probability that exactly 3 out of 6 coin tosses will come up heads?

  (c) The probability of an individual rivet on an airplane wing being defective is 0.004. There are 123 rivets along a seam. What is the probability that 2 or more rivets are defective?

  (d) The probability of finding an individual four leaf clover is $1/10000$. A patch containing 50000 clovers is collected. What is the probability that 7 or more clovers in the patch have four leaves?

3. A new experimental drug is being tested in a clinical trial. The drug is given to 110 patients with a certain illness, and 75 of them show signs of improvement. A placebo (fake drug) is given to a separate control group of 110 patients, and 53 of them show improvement.

  (a) Calculate the measured probability that a patient will show signs of improvement, given that they received the drug. (Express your answer in the form $P(I|D) = P \pm U_P$ with the correct number of significant digits.)

  (b) Calculate the measured probability that a patient will show signs of improvement, given that they did not receive the real drug (placebo). (Express your answer in the form $P(I|D') = P \pm U_P$ with the correct number of significant digits.)

  (c) Regulators require "2σ confidence" for the drug to be approved. That is, the difference between the $P(I|D)$ and $P(I|D')$ must be at least two times greater than the largest uncertainty. Does the drug meet this criteria?

28 Continuous Probability Distributions

At the end of the previous lecture, we saw how a continuous variable, such as electrical resistance, can be mapped to a discrete probability distribution using the bins of a histogram. In this lecture, we will present a new mathematical tool for calculating the probability of observing a continuous variable within any desired range of values.

28.1 The Probability Density Function (PDF)

The **probability density function (PDF)** $\rho(x)$ gives the probability per unit x of observing a value within an infinitesimally small range, between x and $x + dx$. The probability of observing a value x between a and b is calculated by integrating the PDF

$$P(a < x < b) = \int_a^b \rho(x)dx. \tag{28.1}$$

Illustrated in Fig. 28.1, we can see that this probability is represented by the area under the PDF curve $\rho(x)$.

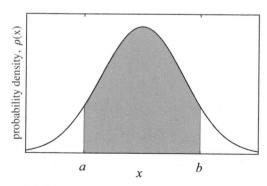

Figure 28.1: The probability of observing a value x between a and b is given by the area under the PDF curve.

28.2 *Normalization, Expected Value, and Standard Deviation*

The continuous probability density function is merely just an extension of the discrete probability distribution into a continuous domain. Thus it shares many of the same properties listed below.

- The probability density function is **normalized** if

$$\int_{-\infty}^{\infty} \rho(x)dx = 1. \qquad (28.2)$$

    In other words, there is a 100% probability of observing x between $-\infty$ and ∞.

- The **mean** or **expected value** is given by the formula

$$\langle x \rangle = \int_{-\infty}^{\infty} x\rho(x)dx. \qquad (28.3)$$

    The expected value can be denoted as $\langle x \rangle$ or $\overline{x}$.

- The **variance** s^2 given by the formula

$$s^2 = \int_{-\infty}^{\infty} (x - \langle x \rangle)^2 \rho(x)dx \qquad (28.4)$$

    The variance of the distribution is simply the **standard deviation** squared s^2. To obtain the standard deviation, we must take the square root of Eq. (28.4). The standard deviation can be denoted as s or σ.

Example 28.1:
Consider the rectangular or "uniform" PDF shown in the plot below.

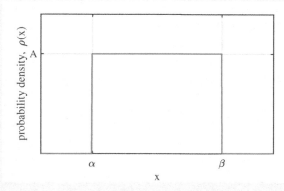

The PDF is given by the analytic piecewise formula

$$\rho(t) = \begin{cases} 0 & \text{if } x < \alpha \\ A & \text{if } \alpha \leq x \leq \beta \\ 0 & \text{if } x > \beta \end{cases}.$$

a) Derive a formula for A in terms of α and β that properly normalizes the PDF.

Solution: Substituting into Eq. (28.2) gives us
$\int_{-\infty}^{\infty} \rho(x)dx = \int_{\alpha}^{\beta} A\,dx = 1$

Evaluating the integral gives us $A(\beta - \alpha) = 1$.

Solving for A, we have $\boxed{A = \dfrac{1}{\beta - \alpha}}$.

b) Derive a formula for the theoretical mean $\langle x \rangle$ in terms of α and β.

Solution: Substituting into Eq. (28.3) gives us
$\langle x \rangle = \int_{-\infty}^{\infty} x\rho(x)dx = \int_{\alpha}^{\beta} xA\,dx = \int_{\alpha}^{\beta} \frac{x}{\beta-\alpha}dx$

Evaluating the integral gives us
$\langle x \rangle = \frac{\beta^2-\alpha^2}{2(\beta-\alpha)} = \frac{(\beta+\alpha)(\beta-\alpha)}{2(\beta-\alpha)}$

$$\boxed{\langle x \rangle = \frac{1}{2}(\beta + \alpha)}$$

c) Derive a formula for the theoretical variance s^2 in terms of α and β.

Solution: Substituting into Eq. (28.3) gives us
$s^2 = \int_{-\infty}^{\infty} (x - \langle x \rangle)^2 \rho(x)dx = \int_{\alpha}^{\beta} \frac{(x-\langle x \rangle)^2}{\beta-\alpha}dx$

Evaluating the integral gives us $s^2 = \frac{(\beta-\langle x \rangle)^3 - (\alpha-\langle x \rangle)^3}{3(\beta-\alpha)}$.

Substituting in our answer from part b), we have $s^2 = \frac{(\beta-\alpha)^3 - (\alpha-\beta)^3}{24(\beta-\alpha)}$,

which can be simplified to $\boxed{s^2 = \dfrac{(\beta - \alpha)^2}{12}}$.

28.3   Cumulative Distribution Function (CDF)

The **cumulative distribution function (CDF)** $F(y)$ gives us the probability that x is less than some value y. Another way to state this is the probability that x is between $-\infty$ and y. To calculate this probability, we use the integral formula from Eq. (28.1)

$$F(y) = P(-\infty < x < y) = \int_{-\infty}^{y} \rho(x)dx. \qquad (28.5)$$

Essentially, it is the are under the PDF curve from $-\infty$ up to some specified point y. Importantly, the CDF always begins at zero for $y \to -\infty$, and goes to one as $y \to +\infty$.

The following example illustrates the properties of the CDF.

Example 28.3:
Consider the rectangular or "uniform" PDF from Example 28.1.

$$\rho(x) = \begin{cases} 0 & \text{if } x < \alpha \\ A & \text{if } \alpha \geq x \leq \beta \, . \\ 0 & \text{if } x > \beta \end{cases}$$

Derive a formula for the cumulative distribution function $F(y)$.

Solution: We will integrate the function piece-by-piece.

- For $y < \alpha$, the PDF $\rho(x) = 0$, so the integral $F(y) = 0$.

- For $\alpha \leq y \leq \beta$, this is where all of the action occurs, so we expect the integral to go from zero to one over this range.
  $F(y) = P(-\infty < x < y) = \int_{\alpha}^{y} \frac{1}{\beta-\alpha}dx.$
  Evaluating the integral gives us $F(y) = \frac{y-\alpha}{\beta-\alpha}$.

- For $y > \beta$, the PDF $\rho(x) = 0$ again, so the integral remains constant from $y > \beta$, such that $F(y) = 1$.

The CDF can be written as a piecewise function.

$$F(y) = \begin{cases} 0 & \text{if } y < \alpha \\ \frac{y-\alpha}{\beta-\alpha} & \text{if } \alpha \leq y \leq \beta \\ 1 & \text{if } y > \beta \end{cases}.$$

A plot of the uniform PDF and its CDF are shown below.

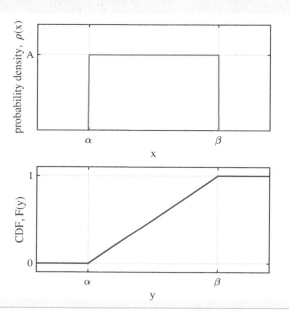

Exercises 28:

Figure 28.2: Chip the alligator.

1. There is a 15 foot long alligator named Chip who lives in a 5 mile long canal in Florida. The probability of finding Chip anywhere along the 5 miles of the canal is given by the uniform probability distribution from Example 28.1, where $\alpha = 0$ miles and $\beta = 5$ miles.

    (a) Compute the normalization constant A in units of 1/mile.

    (b) Sketch the distribution and put labels and important values on the axes. Shade the area under the curve representing the probability of finding chip between mile 3 and 4.

    (c) Compute the theoretical mean, or expected value, for Chip's location. (Don't forget to include units.)

    (d) Compute the theoretical variance in Chip's location. (Don't forget to include units.)

    (e) Compute the theoretical standard deviation in Chip's location. (Don't forget to include units.)

2. Consider the parabola distribution given by the probability density function

$$\rho(x) = \begin{cases} 0 & \text{if } x < -\alpha \\ A(\alpha^2 - x^2) & \text{if } -\alpha \leq x \leq \alpha \\ 0 & \text{if } x > \alpha \end{cases}$$

.

(a) Sketch the distribution with labels for $-\alpha$ and α on the x-axis.

(b) Derive a formula for the normalization constant A in terms of α.

(c) Derive a formula for the theoretical mean, or expected value, $\langle x \rangle$ in terms of α.

(d) Derive a formula for the theoretical variance s^2 in terms of α.

(e) Derive a formula for the cumulative distribution function (CDF).

3. The human population has grown exponentially for the last 2000 years with a growth rate $r = 0.011/\text{year}$. When properly normalized, the birthrate can be interpreted as the probability of someone being born at a given instant in time. Thus, the probability density function for a human being born at a given time is

$$\rho(t) = \begin{cases} Ae^{rt} & \text{if } t < \text{current year (A.D.)} \\ 0 & \text{if } t > \text{current year (A.D.)}, \end{cases}$$

where A is the normalization constant. (The distribution only covers people who have already been born and does not consider people born in the future. Thus $\rho(t) = 0$ for future times.)

(a) Calculate the normalization constant A in units of $(\text{years})^{-1}$.

(b) Sketch the distribution from 0 A.D. to the present year. Shade the area under the curve representing the probability of being born in the middle ages between 500 A.D. and 1500 A.D.

(c) Calculate probability of being born in the middle ages between 500A.D. and 1500A.D.

(d) Calculate probability of being born before the industrial revolution in 1800A.D.

(e) Calculate probability of being born after the industrial revolution in 1800A.D.

(f) Look up the birthday of Guy Fieri. Calculate the probability of being born at some point during Guy Fieri's lifetime.

(g) Why were you born during Guy Fieri's lifetime and not during the middle ages?

29 The Gaussian Distribution

In this lecture, we will present a very specific type of probability density function known as the **Gaussian Distribution** or **Normal Distribution**,

$$\rho(x) = \frac{1}{\sigma\sqrt{2\pi}} \exp\left[\frac{-(x-\bar{x})^2}{2\sigma^2}\right], \tag{29.1}$$

where $\bar{x}$ is the mean and σ is the standard deviation. The Gaussian PDF is plotted below in Fig. 29.1.

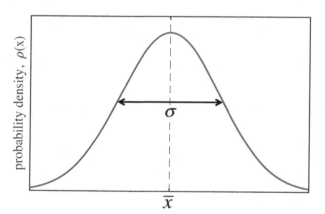

Figure 29.1: The Gaussian or "Normal" PDF is symmetric about its mean value $\bar{x}$ and has a width proportional to its standard deviation σ.

29.1 The Central Limit Theorem

The **central limit theorem** states that the sum of N random numbers, or the net effect of many random processes, will result in a Gaussian distribution that is shaped like a bell curve. A common example of this is rolling dice. The roll of a single fair die yields a uniform probability distribution, as we saw in Lecture 27. Rolling a pair of two dice and taking the sum results in a much different distribution, as shown in Fig. 29.2. Rolling three or four dice and taking the sum results in probability distributions that begin to take on a characteristic bell shape.

According to the central limit theorem, if we were to continue increasing the number of dice, the probability distribution should converge to a continuous Gaussian PDF.

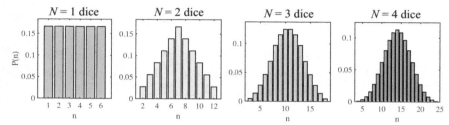

Figure 29.2: The probability distribution for the sum of N dice being rolled together converges toward a Gaussian distribution as N is increased. This is an example of the central limit theorem.

The central limit theorem applies to many other situations beyond dice rolls. Consider, for example, a manufacturing process. Many small random variations occur when fabricating an individual part. The sum total effect of these random variations results in a batch of parts with a Gaussian distribution. For example, the diameter of steel screws or the mass of sugar packets coming out of a factory will follow a Gaussian distribution.

There are two particular instances where the central limit theorem does *not* apply, and the distribution will *not* take on the characteristic Gaussian shape.

- There are not enough "degrees of freedom", or random components, contributing to the distribution. (For example, the pair of only two dice in Fig. 29.2.) In these cases, the Student's t distribution is usually more accurate, and it is part of the reason we used the Student's t-factor when calculating the repeatability uncertainty U_R.

- The values are bounded by some constraint. For example, the diameter of tree trunks cannot be negative, or engineering students cannot score above 100% on an exam. In these cases, it is more appropriate to use a log-normal distribution.

Notes: The Student's t and log-normal distributions are beyond the scope of this book. For more information, please see *Measurements and Data Analysis for Engineering and Science* by Patrick F. Dunn.

29.2   The Gaussian PDF and CDF

The Gaussian PDF given by Eq. (29.1) has a corresponding cumulative distribution function (CDF) given by the formula

$$F(y) = P(x < y) = \frac{1}{2}\left[1 + \text{erf}\left(\frac{y - \bar{x}}{\sigma\sqrt{2}}\right)\right], \qquad (29.2)$$

where erf() is known as the **error function**. The Gaussian PDF and CDF are both shown below in Fig. 29.3. Like any respectable CDF, Equation (29.2) has limits of $F(y \to -\infty) = 0$ and $F(y \to \infty) = 1$.

Notes: The error function is analytically obtained by integrating the Taylor series for Eq. (29.1). For practical purposes, the error function is available as a built-in function in Matlab and most graphing calculators.

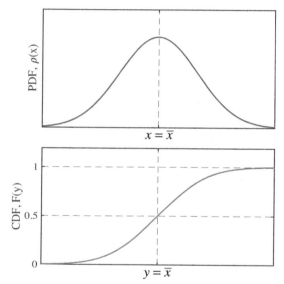

Figure 29.3: The Gaussian PDF (top) and CDF (bottom).

29.3   One-side Error Function Tables (Z-tables)

The Gaussian CDF can be computed using a variety of pre-programmed functions in Matlab or on most graphing calculators. Alternatively, the values can be looked up in a **Z-table**. Inspecting Eq. (29.2), we find the non-dimensional ratio

$$z = \frac{y - \bar{x}}{\sigma}. \qquad (29.3)$$

This parameter tells us how many standard deviations away from the mean a particular value of y is. Found in **Appendix B** at the end of the book, the Z-table lists the area under the curve between the center of the distribution and a particular value of $|z|$. The rows correspond to values of z at one-tenth precision, while the columns correspond to one-hundredth precision. Because the Gaussian is symmetric about the mean, the values are valid for either side of the distribution—positive or negative z.

The following example illustrates how to compute probabilities from a measured mean and standard deviation using either Matlab, a graphing calculator, or a Z-table.

Example 29.1:

The average score on the SAT math exam is $\bar{x} = 580$ with a standard deviation of $\sigma = 80$. The distribution of scores is Gaussian. A student scores 710 on the exam. What percentile is the student in? (What percent of other students scored below 710?)

Solution: For problems like this, the first step is always to sketch the distribution and shade the area under the curve that we are interested in, as shown below.

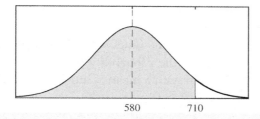

This is a straightforward application of the CDF, where $y = 710$.

$$F(710) = \frac{1}{2}\left[1 + \mathrm{erf}\left(\frac{710 - 580}{80\sqrt{2}}\right)\right]$$

This formula can be typed into Matlab using the erf() function.

```
>> 0.5*(1+erf((710-580)/(80*sqrt(2))))

ans =

   9.4792e-01
```

We can also use the normcdf() function in Matlab.

```
>> normcdf(710,580,80)

ans =

   9.4792e-01
```

Alternatively, we can use the Z-table in **Appendix B**. Plugging in values, we compute $z = \frac{710-580}{80} = 1.63$. The Z-table only gives us one side of the distribution, the area under the curve from $z = 0$ to $z = 1.63$, so we must split up the distribution as shown below.

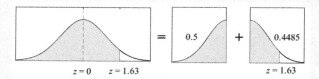

The Gaussian is symmetric, so the area under the curve to the left of $\bar{x}$ or $z = 0$ is 0.5. Using the Z-table, we go down to the row for 1.6 and over to the column for 0.03, and the value for the shaded area is 0.44845. Adding up the area of both sides gives us $P(x < 710) = 0.5 + 0.44845 = 0.94845$.

Lastly, we can use a graphing calculator to compute the probability. The TI-89 calculator has a statistics package that can be installed as a "Flash App". The function "TIStat.normCdf($a,b,\bar{x},\sigma$)" can be used to compute the probability of x between a and b for a Gaussian distribution with a mean $\bar{x}$ and standard deviation σ. TIStat.normCdf($-\infty$,710,580,80) = 0.9479

We see that all three approaches give us approximately 94.8%. Thus, we say that the student is in the 94th percentile.

29.4   Additional Notes on the Gaussian Distribution

The CDF gives us the probability that $x < y$. However, we are often interested in the probability of finding x within a certain range of values $a < x < b$. This can be computed using the formula

$$P(a < x < b) = F(b) - F(a). \tag{29.4}$$

This mathematical operation is illustrated in Fig. 29.4, where we subtract off the tail on the left, whose area is given by $F(a)$.

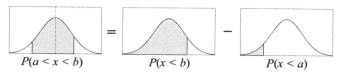

Figure 29.4: To calculate the area under curve between a and b, we simply subtract $F(b) - F(a)$.

Example 29.2:

The average score on the SAT math exam is $\bar{x} = 580$ with a standard deviation of $\sigma = 80$. The distribution of scores is Gaussian. What percent of students will score between 450 and 680?

Solution: Again, the first step is always to sketch the distribution and shade the area under the curve that we are interested in, as shown below.

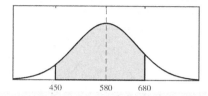

Equation (29.4) can be used to determine the area under the curve, $P(450 < x < 680) = F(680) - F(450)$, which can be computed using the normcdf() function in Matlab.

```
>> normcdf(680,580,80)-normcdf(450,580,80)

ans =

   8.4227e-01
```

Thus, we expect $\boxed{84.2\%}$ of students to score between 450 and 680.

Similar to the previous example, this problem can also be solved using a **Z-table**. Since the Z-table only gives us the area under one side of the distribution, we must split up the shaded region, as illustrated below.

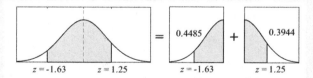

For our values of z, we compute $z = \frac{450-580}{80} = -1.63$ and $z = \frac{680-580}{80} = 1.25$. Looking in Appendix B, we have 0.44845 for $|z| = 1.63$, and 0.39435 for $|z| = 1.25$.

Adding these values gives us $\boxed{0.8428}$ or $\boxed{84.28\%}$.

The Gaussian distribution is also used as a benchmark for standards in science and engineering. Shown in Fig. 29.5, we see that $\approx 68\%$ of values will be within one standard deviation of the mean, and $\approx 95\%$ of values will be within two standard deviations of the mean. This is often referred to as "1σ" and "2σ confidence". The different percentages corresponding 1σ, 2σ, and beyond are listed in Table 29.1.

Notes: Every engineer should memorize these values:

- $\approx 68\%$ are within 1σ of the mean.

- $\approx 95\%$ are within 2σ of the mean.

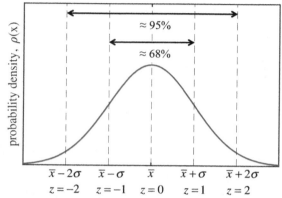

Figure 29.5: For the Gaussian distribution, $\approx 68\%$ of values will be within one standard deviation of the mean, and $\approx 95\%$ of values will be within two standard deviations of the mean..

Table 29.1: These "2-sided" error function values denote percent of values expected to within n standard deviations of the mean.

$n\sigma$	$P[(\bar{x} - n\sigma) < x < (\bar{x} + n\sigma)]$
1σ	68.27%
2σ	95.45%
3σ	99.73%
4σ	99.9937%
5σ	99.999943%
6σ	99.9999998%

Notes: A "Six-sigma (6σ)" manufacturing process is considered to be the gold standard of manufacturing consulting.

As we have mentioned before, no two manufactured parts are ever identical, rather their sizes and features are randomly scattered with some characteristic distribution. By virtue of the central limit theorem, this distribution is typically Gaussian. Manufacturing processes can be dramatically improved through the understanding and manipulation of probability and statistics. This has lead to the lucrative field of **manufacturing consulting**—a career that employs many engineering graduates. The following example demonstrates the power of statistical analysis for improving a manufacturing process.

Example 29.3:

A factory is making extruded metal rods. The rods have a specified diameter of $D = 1.500'' \pm 0.005''$. The tolerance of $\pm 0.005''$ means that a rod's diameter is guaranteed to be in the range of $1.495'' < D < 1.505''$. Any rod with a diameter outside of this range cannot be sold to customers and must be discarded.

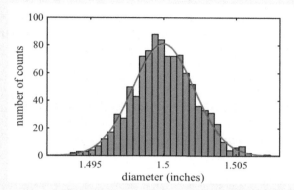

A batch of $N = 1000$ rods is taken from the assembly line and measured, resulting in a mean of $\overline{D} = 1.500''$ and standard deviation $s = 0.002''$. The distribution of diameters, shown in the histogram above, is assumed to be Gaussian.

Solution: As usual, we begin by sketching the distribution and shading the area under the curve. The shaded area is very small, so we expect only a small percentage to be out of spec.

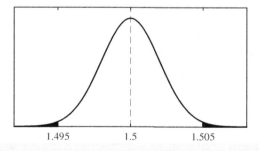

We will use the Z-table for this part of the problem. Calculating z, we have $z = (1.505 - 1.500)/0.002 = 2.5$ for the right half, and $z = (1.495 - 1.500)/0.002 = -2.5$ for the left half.

In the Z-table in Appendix B, we go down to the row for $|z| = 2.5$ and the column for 0.00. The value for one side is 0.49379.

Because the Gaussian is symmetric, we simply multiply this by 2 to give us the area of the un-shaded portion in the middle, 0.9876. The Gaussian is normalized, so its total area is 1. To get the shaded area of the "tails", we subtract $1 - 0.9876 = 0.0124$.

Thus, we expect $\boxed{1.24\%}$ of parts to be out of spec.

b) The company makes 1 million parts per month at a cost of $1.20 per part. How much money will they lose per month due to parts that are out of spec?

Solution: We simply multiply the givens together with appropriate units and take care that the units cancel out correctly.

$$L = \left(\frac{10^6 \text{ opportunities}}{\text{month}}\right)\left(\frac{\$1.20}{\text{defect}}\right)\left(\frac{1.24 \text{ defects}}{100 \text{ opportunities}}\right)$$

$$\boxed{\approx \$14,900 \text{ per month}}$$

c) The factory managers decide that they are losing way too much money on defective parts, so they decide to hire consultants to improve the manufacturing process. The consultants promise to make it a "4σ process". This means that they will decrease the standard deviation from $s = 0.002"$ to $s' = 0.00125"$, such that $z = 0.005/s' = 4$.

Compute the percentage of parts that will be out of spec. and how much money lost per month due to parts that are out of spec. after the process has been improved to 4σ.

Solution: The new distribution is sketched below (dashed line) with the old distribution plotted on top (solid line). The shaded area is now too small to be visible on this scale.

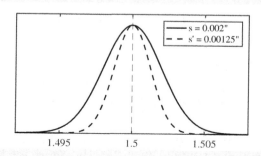

Looking at the Z-table, we see that the values end at $z = 4$, and they are rounded up to 0.49997. To compute this with enough precision, we will need to use a graphing calculator. On a TI-89, we enter the following.

TIStat.normCdf($-\infty$, 1.495, 1.5, 0.00125)
+ TIStat.normCdf(1.505, ∞, 1.5, 0.00125) = 0.000063 or $\boxed{0.0063\%}$

To calculate the monetary losses, simply multiply the givens together with appropriate units and take care that the units cancel out correctly.

$$L = \left(\frac{10^6 \text{ opportunities}}{\text{month}}\right)\left(\frac{\$1.20}{\text{defect}}\right)\left(\frac{0.0063 \text{ defects}}{100 \text{ opportunities}}\right) \approx \boxed{\$76 \text{ per month}}$$

This is a remarkable result! Simply decreasing the standard deviation from 0.002″ to 0.00125″ (a change of only 0.00075″) results in over $14,000 per month in savings! Now, you can see why manufacturing consulting is such a lucrative business.

Notes: In manufacturing consulting, every part coming off the assembly line is called an "opportunity", because it is an opportunity to make money. That opportunity is lost if the part is defective.

Exercises 29:

Assume a Gaussian distribution for all of the following problems.

1. A coastal city is prone to flooding from the "storm surge" of a tropical storm. (Storm surge is a local rise in sea level due to the low air pressure of the storm.) Historical data shows that storm surges cause the water to rise by an average amount $\bar{x} = 15$ feet with a standard deviation of $s = 2.3$ feet.

   (a) The city is 20 feet above normal sea level. What percentage of storm surges will reach the city?

   (b) Tropical storms occur about once a year in this location. Estimate the average amount of time between storms that reach the city.

   (c) Sea levels have been gradually rising over the past 30 years. At the current rate of sea level rise, the city will be 19 feet above sea level at the end of the century. What percentage of storm surges will reach the city at the end of the century?

   (d) Assume that tropical storms still only occur about once a year in this location. Estimate the average amount of time between storms that flood the city at the end of the century.

2. A manufacturing company requires that a certain part have a diameter 1.9000" with a tolerance of ± 0.0020". One thousand of the parts are randomly sampled and inspected, yielding a mean diameter of 1.9003" with a standard deviation of 0.0011".

   (a) If the distribution of part diameters is Gaussian, what percentage of them will be out of spec.?

   (b) The company makes a batch of 10^5 parts per month at a cost of $4.54 per part. How much money will they lose per month by throwing away parts that are out of spec.?

   (c) Consultants are able to refine the manufacturing process such that the new mean diameter is spot-on at 1.9000". What should the new standard deviation be to make the distribution within 4σ of the tolerance?

   (d) Calculate the amount of money the company will lose per month with the refined process that has a mean diameter of 1.9000" and a standard deviation, such that the tolerance is at "4σ".

3. A car company collects data on the lifetime of the transmission in a certain type of car. The average lifetime, or mileage when the transmission fails, is 126,500 miles with a standard deviation of 22,090 miles.

   (a) Use the mean and standard deviation to calculate the probability that the transmission will fail before 100,000 miles.

   (b) The car manufacturer sells a 100,000 mile warrantee on the transmission. If it costs $2,500 to replace the transmission, what is the minimum they should charge for the warrantee for them to expect to break even?

30 Ethical Handling of Data

30.1 Fabricating, Altering, and Withholding Data

Scientists and Engineers have given the world many powerful ideas and inventions that have greatly improved the lives of billions of people, and the very existence of our modern society relies heavily on this work. The incredible power of science and engineering also comes with a great deal of **ethical responsibility**.

So what is our ethical responsibility? We answered this at the beginning of the very first lecture:

"As collegiate scholars, our overall goal is to determine the **truth** and distinguish fact from opinion. As engineers, our overall goal is to design and build products that work well and are safe for people to use."

This may seem straightforward in an academic setting, but in the profit-driven world of business, there is always the temptation to cut corners and break rules for the sake of profit. (Some recent examples of this include the Volkswagen "Dieselgate" scandal, Purdue Pharma's highly addictive OxyContin drug, and the fraudulent Theranos corporation).

Most malfeasances involving scientific data fall into one of three categories:

- **Fabricating Data** – You cannot create false data and say it was measured in an experiment. (Hopefully, this is obvious.)

- **Altering Data** – You cannot change measured data simply to support your hypothesis or make the results appear more favorable to your ultimate goals.

- **Withholding Data** – You cannot delete or disregard measured data, simply because it does not support your ultimate goals.

30.2   Legitimate Reasons for Altering or Withholding Data

There are a few scenarios where it is acceptable to alter or withhold data.

- **Correcting for a "Systematic Error"** – Often times, an experimental method will yield a result that is systematically skewed by some known physical effect. If you truly understand the physical effect that is skewing the data, then it is acceptable to recalibrate or correct the data.

  For example, in Lecture 21.3, we saw that data from a temperature probe can be skewed due to its thermal response time. After determining the thermal time constant, one can then correct the data, as demonstrated in Example 21.1.

- **Equipment Malfunction** – If you believe your instrumentation is not behaving properly, then it is acceptable to withhold or disregard the data it collected. If you choose to do this, you must keep a written record of it in your lab notebook, describing the malfunction and stating the time and date.

- **True Statistical Outliers** – When repeating an experiment N times, it is common to occasionally measure a value that is completely inconsistent with the others for unknown reasons. This errant data point is known as a **statistical outlier**.

  A data point can be classified as a statistical outlier and disregarded if it passes **Grubbs' Test**. Consider the set of N data points in Fig. 30.1, where the data point $x_{outlier}$ is completely inconsistent with the rest. This data point can be removed from the data set if it meets the following criterion

$$\frac{|\bar{x} - x_{outlier}|}{s} > T_{N,\alpha} \tag{30.1}$$

  where $\bar{x}$ is the mean, s is the standard deviation, and $T_{N,\alpha}$ is the critical value for Grubbs' test for N data points with a significance level of α.

Notes: If one chooses to publish corrected data, they must be very explicit and carefully explain how and why they are correcting the data.

Notes: In the formal language of mathematics, we would say our "hypothesis" is that $x_{outlier}$ is an outlier, and our hypothesis is true if Eq. (30.1) holds. This can become confusing to scientists and engineers whose hypotheses normally involve physical phenomena.

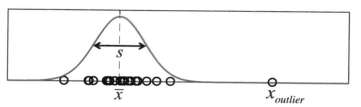

Figure 30.1: A number of data points N are clustered about a mean value $\bar{x}$. One of the data points, denoted as $x_{outlier}$, is a true statistical outlier if Eq. (30.1) holds.

It is common to use a significance level $\alpha = 5\%$, which means that there is only a 5% chance that the outlier was truly sampled from a Gaussian distribution with mean $\bar{x}$ and standard deviation s. The critical values $T_{N,\alpha}$ are computed from Student's t-distribution. However, computing the critical value is usually unnecessary, and they are typically found in a table or come pre-programmed into most statistical software packages, including Matlab.

Table 30.1 lists values of $T_{N,\alpha}$ for different values of N at a significance of $\alpha = 5\%$. The following example shows how to use Grubbs' test using both the tabulated critical values and a pre-programmed function in Matlab.

Table 30.1: Critical values for Grubbs' test for an $\alpha = 5\%$ significance level.

N	$T_{N,\alpha}$
3	1.15
4	1.46
5	1.67
6	1.82
7	1.94
8	2.03
9	2.11
10	2.18
11	2.23
12	2.29
13	2.33
14	2.37
15	2.41
16	2.44
17	2.47
18	2.50
19	2.53
20	2.56

Example 30.1:

An experiment is performed $N = 8$ times, and the results are shown in the adjacent table. For some unknown reason, the last data point is quite a bit off from the average. Use Grubbs' test with a significance level of $\alpha = 5\%$ to determine if the data point is a true statistical outlier.

Solution: According to Table 30.1, $T_{8,5\%} = 2.03$. We then compute the mean and standard deviation, which are determined to be $\bar{x} = 2.17$ and $s = 0.93$. Using Eq. (30.1), we find that $|\bar{x} - x_{outlier}|/s = 2.41 > 2.03$. Thus, it is a true statistical outlier.

Alternatively, we can use the isoutlier() function in Matlab.

Matlab Script:

```
clc
close all

x=[2.15,1.86,1.51,1.84,1.64,2.12,1.82,4.40];
A=isoutlier(x,'grubbs','ThresholdFactor',0.05)
```

Matlab Output:

```
A =

  1×8 logical array

  0  0  0  0  0  0  0  1
```

The "1" in the last position of the output array confirms that the last data point was a true statistical outlier.

Table 30.2: Data for Example 30.1.

data, x
2.15
1.86
1.51
1.84
1.64
2.12
1.82
4.40

Exercises 30:

1. Consider the following scenarios and explain what is the most ethical and appropriate course of action.

   (a) An engineer has spent the entire day testing components for a jet aircraft. At the end of the day, the data she has collected does not seem consistent with what she observed the previous week. Then, she notices there is a little red light flashing on the instrumentation panel.

   (b) A biomedical research scientist is testing a new drug in a clinical trial. After rigorous statistical analysis of the results, the drug does not appear to be any more effective than a simple placebo. However, at the quarterly shareholders meeting, the CEO expresses great optimism and tells investors "We are the on the verge of a breakthrough!"

   (c) The owner of a toy company outsources production to a factory in a foreign country. The cheap labor and lax regulations in this country will allow for profits to double. When touring the factory, it is revealed that the paint used on the toys contains toxic chemicals. However, there are no laws saying this type of paint cannot be used on toys.

data, x
3.45
3.16
2.84
7.31
2.99
2.93
3.61
3.02
3.29
3.11
3.32
2.97

2. Consider the data in the adjacent table.

   (a) Use Grubbs' test to identify if there are any statistical outliers. Do this manually using the critical values in Table 30.1.

   (b) Repeat Grubbs' test using the isoutlier() function in Matlab.

31 Sensor Round-Up

It's round up time, y'all! We will now review for the final exam by going over all of the sensors we have covered, plus a few more that we have not yet covered.

Notes: The parameter being measured is *italicized* in all of the following definitions.

31.1 Temperature Measurements

Resistive Temperature Detector (RTD) – The resistance of a thin metal wire, usually platinum, increases linearly with *temperature*.

Thermistor – The resistance of a ceramic or semiconductor depends on *temperature* via the Steinhart Equation.

Thermocouple – A voltage naturally forms across the junction of two different metals because of the Seebeck effect. This voltage is amplified by a special compensation circuit and used to compute the *temperature*.

Pyrometer – Warm objects emit blackbody radiation. The intensity and shape of the emitted spectrum can be used to deduce the temperature of the object. This allows for "non-contact" *temperature* measurements.

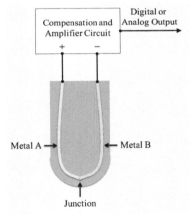

Figure 31.1: A thermocouple consists of two different metals connected at a junction. The Seebeck effect causes a voltage to form across the junction that is related to the temperature.

31.2 Fluid Measurements

Manometer – *Pressure* is measured via the height of a liquid in a tube. The height of a liquid h in a tube is related to the pressure difference Δp at the ends of the tube via the hydrostatic pressure balance $\Delta p = \rho g h$.

Capacitive Pressure Transducer - This sensor uses two diaphragms to form a parallel plate capacitor. Applying a *pressure* displaces the diaphragms and changes the electrical capacitance.

Piezoelectric Pressure Sensor – Applying a *pressure* difference across a piezoelectric crystal causes it to compress or expand, which results in a voltage across the surface.

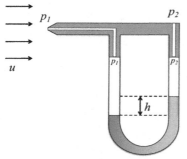

Figure 31.2: An pitot tube is connected to a U-tube manometer

Pitot Static Probe or **Pitot Tube** – Moving air stagnates in the orifice of a tube. The stagnation pressure is used to deduce the *airspeed* via the Bernoulli equation.

Hotwire Anemometer – Current is driven through a thin tungsten wire, causing it to heat up. Airflow over the hot wire then cools it off, and the equilibrium temperature of the wire depends on the airspeed. The wire also behaves as an RTD, so the electrical resistance of the wire can be used to deduce the *airspeed*.

Rotameter - The drag force of fluid flowing through a tapered tube levitates a bead, and the height of the levitating bead is used to determine the *flow rate* (volume/time) of the fluid.

Venturi Tube – A "converging-diverging" nozzle shaped like an hourglass is used to measure *flow rate* (volume/time) via the Bernoulli effect.

Paddle Wheel Flow Meter – Flow through a pipe spins a paddle wheel. The angular speed of the paddle wheel is used to deduce the *flow rate* (volume/time).

Schlieren Visualization – This optical technique allows one to capture an image of *density gradients* in a gas. It is particularly useful when studying supersonic flows moving faster than the speed of sound.

Particle Image Velocimetry (PIV) – Small seed particles are placed in a fluid flow, and the seed particles move with the fluid. Two photographs of the seed particles are taken, separated by a time interval Δt. Comparing the images, one can estimate the change in position of different seed particles $\Delta \vec{x}$, and then compute the velocity of each seed particle $\vec{u} = \Delta \vec{x}/\Delta t$. Importantly, PIV gives you the *velocity at many different locations* in the flow.

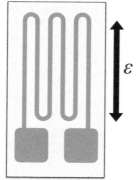

Figure 31.3: A strain gage consists of a thin metal wire that zig-zags back and forth inside of a plastic film.

31.3   Stress-strain Measurements

Strain Gage – A long, thin wire is encapsulated in plastic and glued to a solid surface. The resistance of the strain gage changes as the surface is stretched or compressed, allowing for the *strain* to be measured.

Load Cell – This electronic sensor is used to measure *force*. It typically consists of a well-known elastic metal with strain gages mounted to it. The elastic metal stretches or compresses under the applied load, resulting in an electronic signal from the strain gages.

Linear Variable Displacement Transducer or **Linear Variable Differential Transformer (LVDT)** – The electromagnetic inductance of a solenoid coil of wire changes when a ferromagnetic rod is inserted by a *distance* Δx.

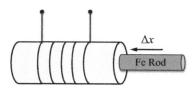

Figure 31.4: An LVDT typically consists of a solenoid with a rod made of some ferromagnetic material.

Extensometer – This electronic sensor can measure extremely small changes in *length* Δx.

Load Frame – This is a machine capable of stretching or compressing solid materials with thousands of pounds of force. It is commonly used to measure *stress-strain properties* of various materials and mechanical parts.

Notes: "Ultrasonic" means the acoustic wave has a frequency $f > 18$ kHz, which is too high pitched for an adult human ear to hear.

31.4 Position, distance, and range

Ultrasonic Range Finder – A piezoelectric transmitter and receiver are mounted adjacent to one another. The transmitter sends out an acoustic "chirp". The chirp echoes off a nearby object and comes back to the receiver. The time between the transmission and reception of the echo Δt can be used, along with the speed of sound $c = 340$ m/s, to deduce the *distance* to the object $\Delta x = c\Delta t/2$.

Light Detection and Ranging (LIDAR) – An infrared laser and photodetector are mounted adjacent to one another. The laser sends out a light pulse. The light pulse reflects off a nearby object and comes back to the photo-sensor. The time between the transmission and reception of the light pulse Δt can be used, along with the speed of light $c = 3 \times 10^8$ m/s, to deduce the *distance* to the object $\Delta x = c\Delta t/2$.

Radio Detection and Ranging (RADAR) - A transmitter tower sends out radio waves, which scatter off nearby objects. The scattered radio waves are measured by receiver towers, and the power, phase, and polarization of the scattered waves can be used to deduces the *position and velocity* of the nearby object.

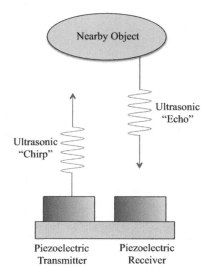

Figure 31.5: An ultrasonic rangefinder consists of a piezoelectric transmitter and receiver.

Notes: GPS receivers have become incredibly inexpensive and can easily be implemented in any student project.

Global Positioning System (GPS) – This is a system of at least 24 satellites orbiting the earth. Each satellite constantly broadcasts its current position and time stamp, and a GPS receiver on earth listens for these broadcasts. Based on the timestamps, the receiver can deduce the distance to each of the satellites using the speed of light. If the receiver knows the position and distance to at least *four* different satellites, it can calculate its own *position on earth* using a mathematical operation called "trilateration".

31.5   Angular Position and Angular Speed

Potentiometer – This device measures the angular position of a rotating shaft or knob using a fixed resistive arc and conductive brush coupled to the shaft. The relative resistance between the brush and the ends of the resistive arc is related to the *angular position* of the knob.

Quadrature Encoder – This digital sensor measures the angular position of a rotating shaft using a slotted wheel and two photogates. The slotted wheel is coupled to the shaft and passes through the photogates, creating digital pulses. A counter circuit counts the digital pulses, which is proportional to the *angular position*.

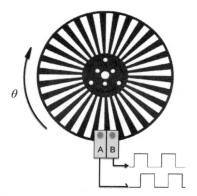

Figure 31.6: A quadrature encoder consists of a slotted wheel mounted on a rotating shaft. The photogates detect the slots in the wheel as the shaft is rotated.

Absolute Encoder - A wheel with a special pattern of holes is coupled to a shaft. As the shaft is rotated, an array of photo-sensors reads the pattern on the wheel as a binary code, which denotes the *angular position* of the shaft.

Tachometer – This sensor measures the *angular speed* of a rotating mechanical part. Tachometers can be mechanical, optical, or magnetic.

31.6   Inertial Measurements

MEMS Accelerometer - This device uses the capacitance of very small inter-digitated electrodes mounted to a spring to measure *acceleration*.

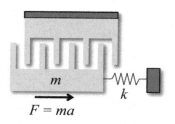

Figure 31.7: The MEMS accelerometer has interdigitated electrodes. One set of electrodes is is connected to a spring and moves, while the other is fixed and does not move.

Mechanical Gyroscope – A spinning flywheel is coupled to an aircraft or spacecraft via a series of nested rings known as "gimbals". To conserve angular momentum, the spinning flywheel maintains its angular position in space, even as the aircraft or spacecraft is rotated. Encoders or potentiometers are used to measure the angles of the gimbals, which typically correspond to pitch, roll, and yaw—the *angular orientation*, or *attitude*, of the craft. The flywheel will also precess when it is subject to a torque or angular acceleration. The rate of precession can be used to determine *angular acceleration*.

Ring Laser Gyroscope (RLG) – Two laser beams, propagating in opposite directions, are reflected around a series of mirrors that form a ring. If the device is rotated in a certain direction, one of the laser beams becomes red-shifted, while the other beam is blue-shifted. This results in an interference pattern that can be measured and used to deduce *angular position* and *angular speed*.

Inertial Measurement Unit (IMU) – This term refers to any device that measures *linear* or *angular acceleration*.

Attitude, Heading, and Reference System (AHRS) – This electronic device is used to determine the *angular position* (attitude), *velocity* (heading), and *location* of an aircraft. It contains a whole suite of sensors and electronics—MEMS accelerometers, ring laser gyroscopes, and GPS receivers.

Notes: The ring laser gyro has replaced old mechanical gyroscopes in most modern aircraft, because they are more accurate and reliable. In particular, the RLG has no moving parts.

Appendices

Appendix A

Student's t-table – Values of $t_{v,\%C}$ represent half the width of the area under the Student's t-distribution that contains an expected percentage of values %C.
(Use these to compute repeatability uncertain U_R.)

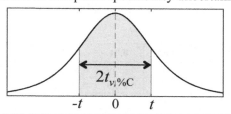

v	$t_{v,50\%}$	$t_{v,68.2\%}$	$t_{v,90\%}$	$t_{v,95\%}$	$t_{v,99\%}$
1	1.000	1.833	6.314	12.706	63.657
2	0.816	1.319	2.920	4.303	9.925
3	0.765	1.195	2.353	3.182	5.841
4	0.741	1.140	2.132	2.776	4.604
5	0.727	1.109	2.015	2.571	4.032
6	0.718	1.089	1.943	2.447	3.707
7	0.711	1.075	1.895	2.365	3.499
8	0.706	1.065	1.860	2.306	3.355
9	0.703	1.057	1.833	2.262	3.250
10	0.700	1.051	1.812	2.228	3.169
11	0.697	1.046	1.796	2.201	3.106
12	0.695	1.042	1.782	2.179	3.055
13	0.694	1.038	1.771	2.160	3.012
14	0.692	1.035	1.761	2.145	2.977
15	0.691	1.033	1.753	2.131	2.947
16	0.690	1.031	1.746	2.120	2.921
17	0.689	1.029	1.740	2.110	2.898
18	0.688	1.027	1.734	2.101	2.878
19	0.688	1.026	1.729	2.093	2.861
20	0.687	1.024	1.725	2.086	2.845
21	0.686	1.023	1.721	2.080	2.831
22	0.686	1.022	1.717	2.074	2.819
23	0.685	1.021	1.714	2.069	2.807
24	0.685	1.020	1.711	2.064	2.797
25	0.684	1.019	1.708	2.060	2.787
30	0.683	1.015	1.697	2.042	2.750
40	0.681	1.011	1.684	2.021	2.704
50	0.679	1.009	1.676	2.009	2.678
100	0.677	1.004	1.660	1.984	2.626
∞	0.674	0.999	1.645	1.960	2.576

NOTE: Student's t-factors can be computed in Matlab for any value of %C and v using the formula tinv(0.5*(1+C),nu), where C is the confidence expressed as a fraction < 1.

Appendix B

Z-table – One-sided error function values for computing the Gaussian CDF

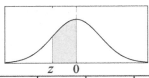 or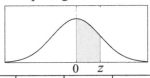

$|z|$	0.00	0.01	0.02	0.03	0.04	0.05	0.06	0.07	0.08	0.09
0.0	0.00000	0.00399	0.00798	0.01197	0.01595	0.01994	0.02392	0.02790	0.03188	0.03586
0.1	0.03983	0.04380	0.04776	0.05172	0.05567	0.05962	0.06356	0.06749	0.07142	0.07535
0.2	0.07926	0.08317	0.08706	0.09095	0.09483	0.09871	0.10257	0.10642	0.11026	0.11409
0.3	0.11791	0.12172	0.12552	0.12930	0.13307	0.13683	0.14058	0.14431	0.14803	0.15173
0.4	0.15542	0.15910	0.16276	0.16640	0.17003	0.17364	0.17724	0.18082	0.18439	0.18793
0.5	0.19146	0.19497	0.19847	0.20194	0.20540	0.20884	0.21226	0.21566	0.21904	0.22240
0.6	0.22575	0.22907	0.23237	0.23565	0.23891	0.24215	0.24537	0.24857	0.25175	0.25490
0.7	0.25804	0.26115	0.26424	0.26730	0.27035	0.27337	0.27637	0.27935	0.28230	0.28524
0.8	0.28814	0.29103	0.29389	0.29673	0.29955	0.30234	0.30511	0.30785	0.31057	0.31327
0.9	0.31594	0.31859	0.32121	0.32381	0.32639	0.32894	0.33147	0.33398	0.33646	0.33891
1.0	0.34134	0.34375	0.34614	0.34849	0.35083	0.35314	0.35543	0.35769	0.35993	0.36214
1.1	0.36433	0.36650	0.36864	0.37076	0.37286	0.37493	0.37698	0.37900	0.38100	0.38298
1.2	0.38493	0.38686	0.38877	0.39065	0.39251	0.39435	0.39617	0.39796	0.39973	0.40147
1.3	0.40320	0.40490	0.40658	0.40824	0.40988	0.41149	0.41309	0.41466	0.41621	0.41774
1.4	0.41924	0.42073	0.42220	0.42364	0.42507	0.42647	0.42785	0.42922	0.43056	0.43189
1.5	0.43319	0.43448	0.43574	0.43699	0.43822	0.43943	0.44062	0.44179	0.44295	0.44408
1.6	0.44520	0.44630	0.44738	0.44845	0.44950	0.45053	0.45154	0.45254	0.45352	0.45449
1.7	0.45543	0.45637	0.45728	0.45818	0.45907	0.45994	0.46080	0.46164	0.46246	0.46327
1.8	0.46407	0.46485	0.46562	0.46638	0.46712	0.46784	0.46856	0.46926	0.46995	0.47062
1.9	0.47128	0.47193	0.47257	0.47320	0.47381	0.47441	0.47500	0.47558	0.47615	0.47670
2.0	0.47725	0.47778	0.47831	0.47882	0.47932	0.47982	0.48030	0.48077	0.48124	0.48169
2.1	0.48214	0.48257	0.48300	0.48341	0.48382	0.48422	0.48461	0.48500	0.48537	0.48574
2.2	0.48610	0.48645	0.48679	0.48713	0.48745	0.48778	0.48809	0.48840	0.48870	0.48899
2.3	0.48928	0.48956	0.48983	0.49010	0.49036	0.49061	0.49086	0.49111	0.49134	0.49158
2.4	0.49180	0.49202	0.49224	0.49245	0.49266	0.49286	0.49305	0.49324	0.49343	0.49361
2.5	0.49379	0.49396	0.49413	0.49430	0.49446	0.49461	0.49477	0.49492	0.49506	0.49520
2.6	0.49534	0.49547	0.49560	0.49573	0.49585	0.49598	0.49609	0.49621	0.49632	0.49643
2.7	0.49653	0.49664	0.49674	0.49683	0.49693	0.49702	0.49711	0.49720	0.49728	0.49736
2.8	0.49744	0.49752	0.49760	0.49767	0.49774	0.49781	0.49788	0.49795	0.49801	0.49807
2.9	0.49813	0.49819	0.49825	0.49831	0.49836	0.49841	0.49846	0.49851	0.49856	0.49861
3.0	0.49865	0.49869	0.49874	0.49878	0.49882	0.49886	0.49889	0.49893	0.49896	0.49900
3.1	0.49903	0.49906	0.49910	0.49913	0.49916	0.49918	0.49921	0.49924	0.49926	0.49929
3.2	0.49931	0.49934	0.49936	0.49938	0.49940	0.49942	0.49944	0.49946	0.49948	0.49950
3.3	0.49952	0.49953	0.49955	0.49957	0.49958	0.49960	0.49961	0.49962	0.49964	0.49965
3.4	0.49966	0.49968	0.49969	0.49970	0.49971	0.49972	0.49973	0.49974	0.49975	0.49976
3.5	0.49977	0.49978	0.49978	0.49979	0.49980	0.49981	0.49981	0.49982	0.49983	0.49983
3.6	0.49984	0.49985	0.49985	0.49986	0.49986	0.49987	0.49987	0.49988	0.49988	0.49989
3.7	0.49989	0.49990	0.49990	0.49990	0.49991	0.49991	0.49992	0.49992	0.49992	0.49992
3.8	0.49993	0.49993	0.49993	0.49994	0.49994	0.49994	0.49994	0.49995	0.49995	0.49995
3.9	0.49995	0.49995	0.49996	0.49996	0.49996	0.49996	0.49996	0.49996	0.49997	0.49997
4.0	0.49997	0.49997	0.49997	0.49997	0.49997	0.49997	0.49998	0.49998	0.49998	0.49998

Index

Made in the USA
Columbia, SC
09 August 2021